RIDING ON AIR

A History of Air Suspension

Jack Gieck, P.E.

SAE INTERNATIONAL®

Society of Automotive Engineers, Inc.
Warrendale, Pa.

Library of Congress Cataloging-in-Publication Data

Gieck, Jack.
Riding on air : a history of air suspension / Jack Gieck.
p. cm.
Includes bibliographical references and index.
ISBN 0-7680-0454-3
1. Motor vehicles—Air suspension—History. I. Title.
TL257.3.G54 1999
629.2'43—dc21 99-33364
CIP

400 Commonwealth Drive
Warrendale, PA 15096-0001 U.S.A.
Phone: (724)776-4841
Fax: (724)776-5760
E-mail: publications@sae.org
http://www.sae.org

ISBN 0-7680-0454-3

SAE Order No. R-235

*To my collaborators on this book, Tom Bank and Art Hirtreiter—
who made air suspension possible*

CONTENTS

PREFACE

Grecian chariots rode hard. As these ancient war machines rumbled over the rugged roads of Babylonia and Egypt, and later raced across the wooden floor of the Roman Colosseum, every minor irregularity encountered by their solid wooden wheels was faithfully transmitted through their solid axles to the bed of the vehicle, some of the bumps imparting several *g*s—which may account for their drivers electing to stand rather than attempting to sit down.

The first recorded efforts to achieve a modicum of ride comfort were literally "suspensions." The passenger compartments of seventeenth century coaches in France and England were suspended by long leather straps attached to four vertical wooden posts anchored to the box frame of the horse-drawn vehicle. The date of conception of this invention is said to be circa 1430, when, in the town of Kocs (from which the word "coach" is derived), in what is now southwestern Czechoslovakia, a royal personage is said to have complained about the rough ride afforded by the wagons in which she was transported.[1]

Eighteenth century stagecoach passengers gratefully accepted the introduction of wooden springs (hickory was the preferred raw material), frequently in combination with leather straps, and later S-shaped iron or steel springs forged by blacksmiths of the time. By 1796, builders of a two-horse English curricle had inserted coil springs as spreaders between the tensioned leather straps to further soften the ride. Eventually, as steel became commercially available in the mid-nineteenth century, multi-leaf elliptical springs appeared on horse-drawn coaches in both England and the United States.[2]

But some of the authors of what passed for the technical literature of the time wrote about the promise offered by "the springiness of air itself" as a potential medium to isolate the passenger from the road. And it wasn't long before the idea appeared in a series of patents.

I first heard of air suspension in 1947, shortly after I joined Firestone, having read Roy W. Brown's SAE technical paper, "Air Springs—Tomorrow's Ride," which he had presented to the annual meeting of the Society of Automotive Engineers a decade earlier.[3] Brown's ingenious but hopelessly complex pneumatic system for passenger cars had generated little enthusiasm in the Depression-era economy of 1936. But, unknown to me at the time, Brown and Firestone engineer Fred Haushalter were already at work with engineers of General Motors Truck & Coach Division in Pontiac, Michigan, on a secret development program that would revolutionize the postwar bus industry and would dominate my own life for more than a decade.

As General Motors began gearing up for production of its new Greyhound buses in 1951, I became personally involved with air springs as a product development engineer in Firestone's Detroit office. For me, it would be the beginning of a sometimes precarious career adventure, including what initially seemed an unqualified success for the technology. Beginning on buses, air suspension expanded into passenger rail vehicles, only to be followed by a crashing failure on passenger cars. But after several precarious years, air suspension began to win almost universal acceptance on trucks and trailers, and then in mass transportation—eventually making a successful return to passenger cars.

After decades of hiatus on passenger automobiles (with the notable exception of Delco *Superlift* shock absorbers in combination with coil springs on the rear of the Cadillac and the El Camino pickup), Ford Motor Company and several European manufacturers would make air springs a practical, reliable suspension system for personal transportation.

During the years I worked on the technology, eventually becoming product manager, I formed lifelong friendships with the engineering associates who were my partners in launching the product, with the customer engineers with whom I worked (and sometimes vigorously argued), and even with the chief engineer of my major competitor.

Without the innovative contributions of my primary collaborators on this book, Tom Bank, Manager of Airide Development for Firestone, and Art Hirtreiter, Goodyear's Chief Engineer for Suspensions and Air Springs—who created manufacturing processes to make reliable elastomeric air springs in scores of sophisticated shapes and constructions for hundreds of diverse applications—the product might have died. Providing us with the incentive to make sure that did not happen on their watch were such customer engineers as John Locklin, John Killinger, Don Manning, and John Hoban. All of the above have contributed generously to the content of this book, as did Tom Burkley, John Grafton, and Andy Palguta of Goodyear; Charlie Slemmons of General; as well as Herb Deist, Jack

Hollis, Dave Thomas, Gary Reynolds and Steve Lindsey of Firestone. Many of the book's vintage illustrations are the product of the invaluable pack-rat tendencies of self-appointed historical archivists John Grafton of Goodyear and Larry Wilson of Firestone.

I am also indebted to George Villec (SAE's reviewer for this book), and to Michael Soltis of Ford Motor Company for their substantial contributions—and of course, to my friends at Firestone Industrial Products, who made a major industry out of the modest business that Fred Haushalter, Tom Bank, Mike Kray, Gerry Marsh, Herb Deist, Bob Weir, Al Boyer, Mary Jean Hosier, and I started half a century ago.

The history of air suspension goes back more than a century and a half, which will become apparent as we review concepts that have periodically emerged to advance the technology. It is often easy to see how one idea led to another. But in offering a chronology of these developments, it is never my intention to suggest that any of these inventors deliberately plagiarized innovations from prior art. Rather, I think we see examples of technological evolution—ideas whose time had inevitably come—just as similar scientific discoveries are often simultaneously announced half a world apart. Many of these innovators might honestly confess, however, as Isaac Newton did in a rare moment of modesty, "If I have seen farther than others, it is because I have stood on the shoulders of giants."

Jack Gieck

Akron, Ohio

Chapter One

THE MEDIUM

Half a century before automobiles emerged on American and European roads, engineers dreamed of floating their vehicles on cushions of air. In 1847, the year Thomas Edison was born, only three years after Charles Goodyear's rubber vulcanization patent was issued,[1] inventor John Lewis was granted U.S. Patent No. 4,965 for "Pneumatic Springs for Railroad Cars, Locomotives, Burden-Cars, Bumpers &c." (see Fig. 1.1).[2]

In his patent, Lewis rhapsodizes about "the perfect elasticity of atmospheric air," admitting that he is not the first to be fascinated with the medium. "[Air] has long since suggested the idea of applying it as a substitute for metal springs," his patent reads, "but this has always been effected by means of a metallic cylinder and piston, which permits the escape of contained air." A sliding piston, he goes on to explain, also introduces friction and wear. Instead, Lewis advocates his concept of metal "respiratory chambers ... connected together by two or more belts of India rubber cloth."

With these words, Lewis claimed his invention of the first rubberized fabric air spring. He may, in fact, have been acquainted with Charles Goodyear, because both lived in New Haven, Connecticut, at the time.

The same year that Lewis's patent was granted, the concept of riding on air proved of interest to a new magazine of the time; but the application it described was not exactly an air *spring*. In its May, 1847, issue, the neonate *Scientific American* reported that:

A number of [horse-drawn] cabs with newly invented wheels have just been put on the road in London. Their novelty consists in the entire absence of springs. A hollow tube of India rubber about a foot in diameter, inflated with air, encircles each wheel in the manner of a tire, and with this simple but novel appendage the vehicle glides noiselessly along, affording the greatest possible amount of cab comfort to the passenger.[3]

A "tire," at the time, was understood to mean a band of *iron* wrapped around the circumference, or rim, of a (wooden) wheel to reduce road wear. The very fat twelve-inch section of this original, truly "balloon" tire suggests that it must have had a very low inflation pressure, if any. It would be almost half a century (1888) before the Scottish

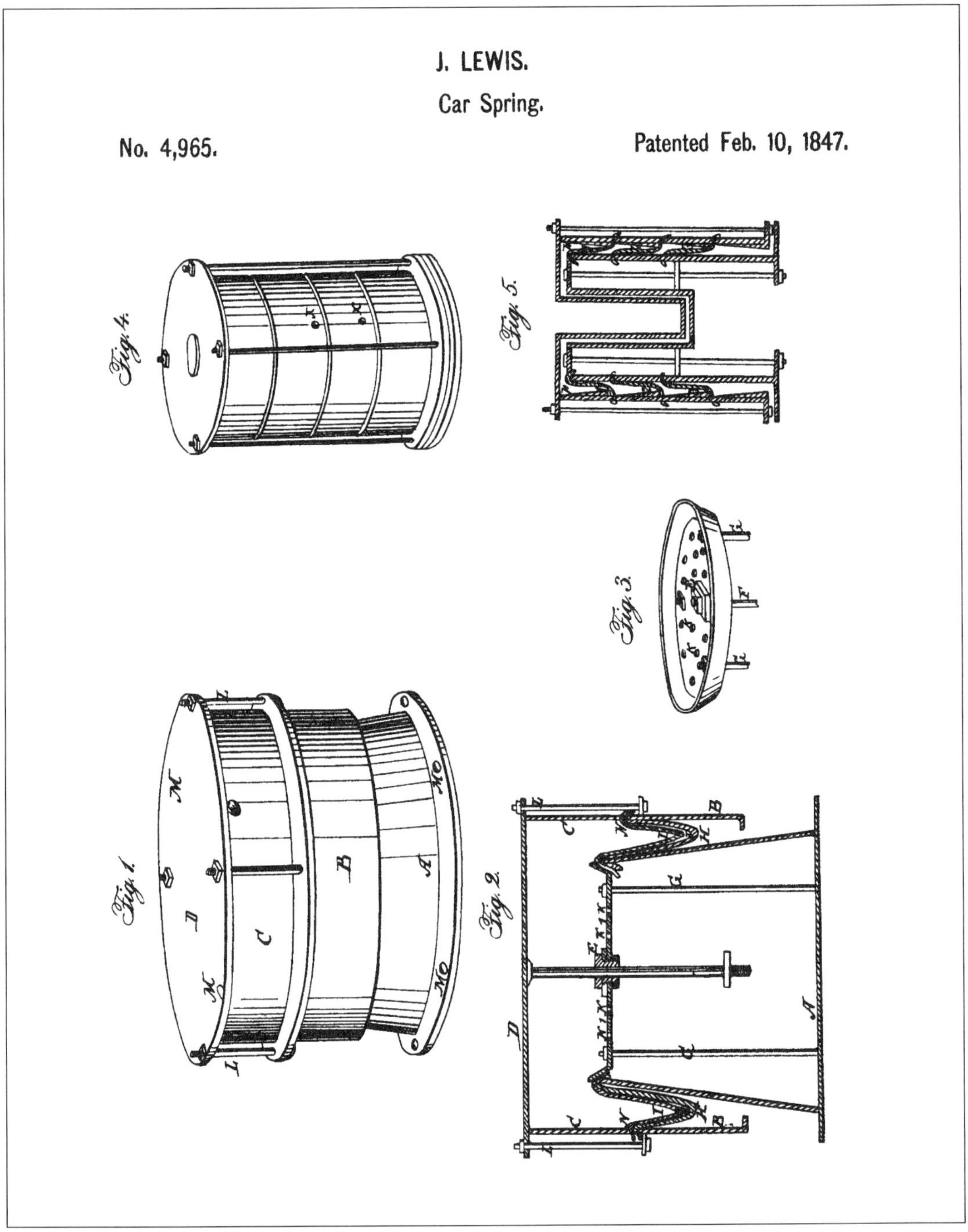

Fig. 1.1 In 1847, John Lewis of New Haven, Connecticut, was issued U.S. Patent 4,965 for the first air spring made of a flexible rubber material—a "Pneumatic Spring for Railroad Cars."

veterinarian, John Boyd Dunlop, would develop a practical pneumatic tire—which he initially intended for bicycles.[4]

Floating a vehicle on cushions of air sounds like a marvelous idea. Air suspension invokes an image of passengers and cargo flying above the road, completely isolated from any bumps on its surface. Indeed, trade journal advertisements announcing the Firestone innovation on General Motors buses in 1953 pictured a Greyhound luxury coach that had sprouted *wings*.

Knowledgeable engineers were not impressed by the ad. Vehicle ride, they knew, is independent of the spring medium. It is a function of the *stiffness* of the springs of the vehicle (whatever they are made of), the *load* resting on them, and any *damping* devices in the suspension, such as hydraulic shock absorbers—*and* the vehicle's tires, of course.

What air suspension really offers is *adjustable*, variable-rate springs whose stiffness can be fine-tuned as the load in the vehicle changes. This unique advantage of air springs—one that would not be realized until the twentieth century—was disclosed in another *Scientific American* article, this one in June, 1861, which reported an invention by I. W. Hoagland of New Brunswick, New Jersey: a "New and Improved Pneumatic Spring for [railway] Carriages," which, the magazine explained, "is easily adjustable to any degree of compression desired" (see Fig.1.2).[5]

"The mechanical engineer who devised these ingenious modifications," the magazine article continues, "is satisfied that he has overcome all the difficulties encountered in the use of the air spring, and ... by the adoption of his improvements, the great advantages of that most perfect of all springs can be practically realized." Hoagland's confidence notwithstanding, those of us who struggled with the first practical applications of air springs on buses and passenger cars in the 1950s would not have agreed.

That Lewis's and Hoagland's dreams were a century ahead of rubber technology is evident in the language of their own patents. Lewis's proposal, with its deliberate redundancy of two, or even three, nesting rubberized fabric diaphragms between his "respiratory chambers" was a hedge against leakage—as was his "interposing alcohol or water" on both sides of the flexible joint, "the more effectually to prevent the escape of the air." Other inventors, e.g., William R. Fee, with his pneumatic "Car Spring" (Fig. 1.3) and George. M. Alsop's "Carriage Spring" (Fig. 1.4) picture "vessels" partially filled with water, oil, or other liquids.[6]

I. W. Hoagland's pneumatic spring for railroad cars, "about nine inches [23 cm] in diameter, and nine inches in height" offered a more complex solution to the "difficulty of preventing air from leaking out, either around the piston or through the walls." The magazine pictured a kind of boot-strap arrangement which captured the energy available in the bouncing of the suspension. In the center of Hoagland's large, primary air spring

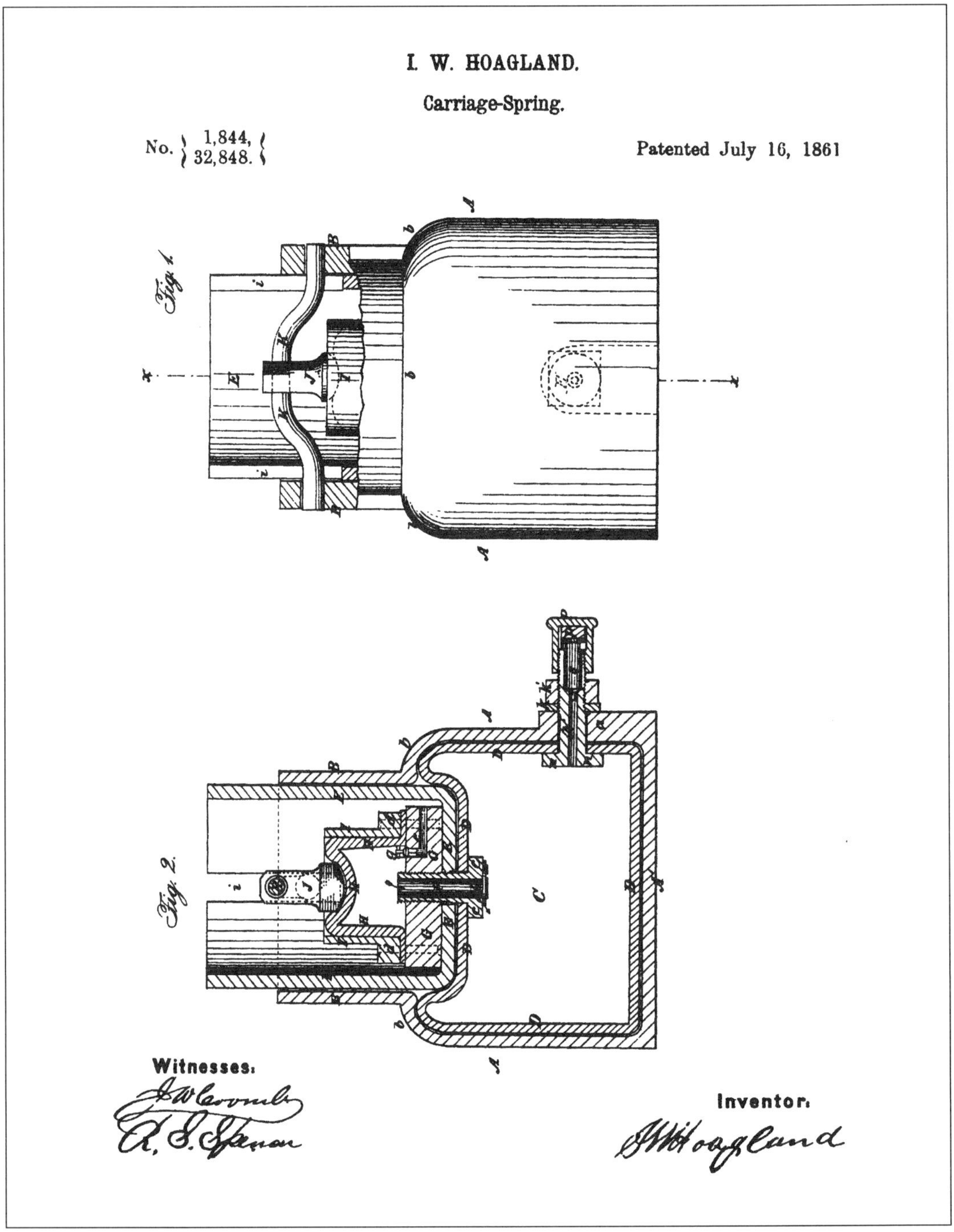

Fig. 1.2 I. W. Hoagland's 1861 air spring patent, also intended for railway carriage suspensions, had its own built-in air compressor to compensate for leakage. The bouncing motion of the suspension pumped a miniature air spring which kept the primary spring inflated.

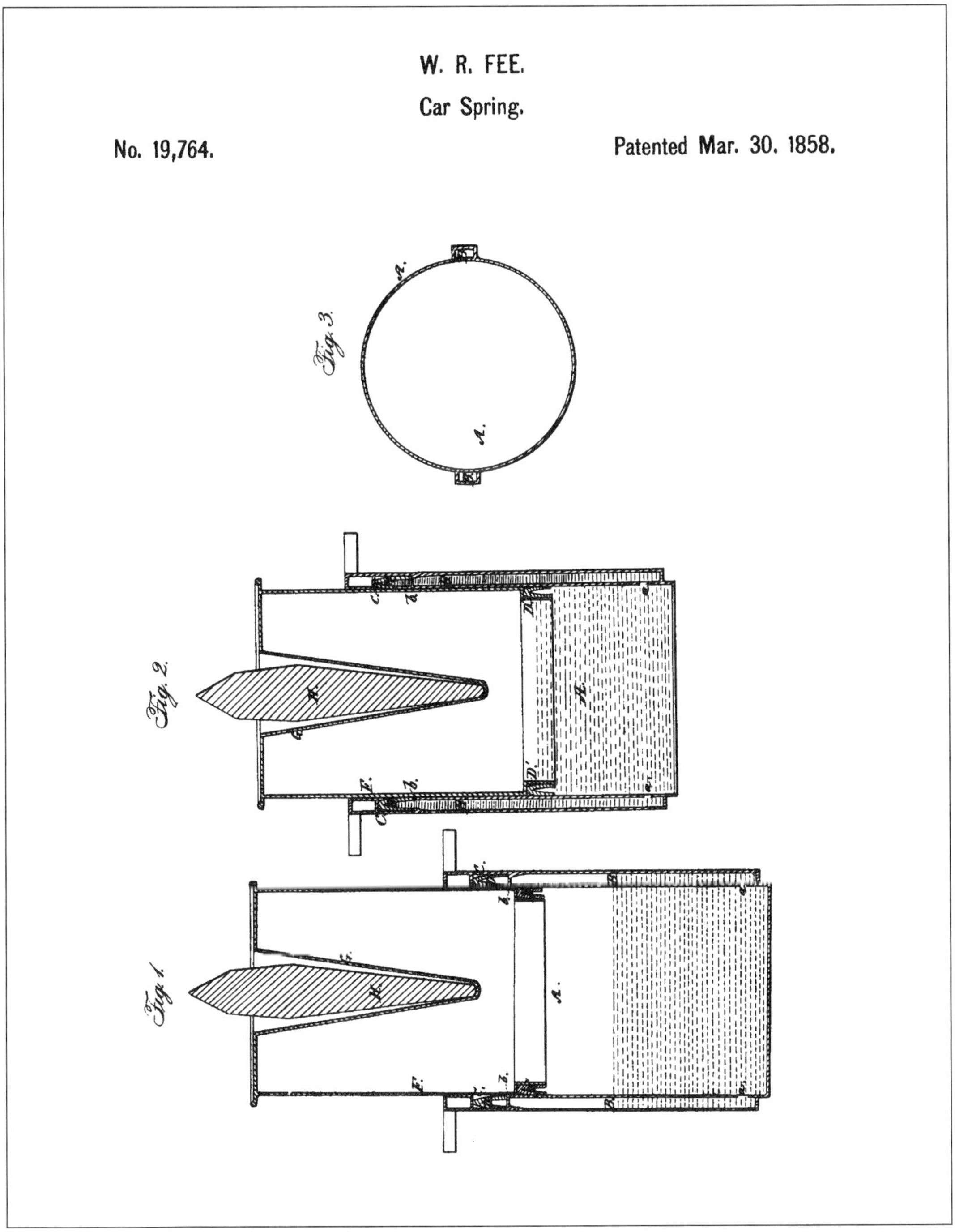

Fig. 1.3 To limit air leakage, William Fee partially filled his 1858 air spring for rail cars with water.

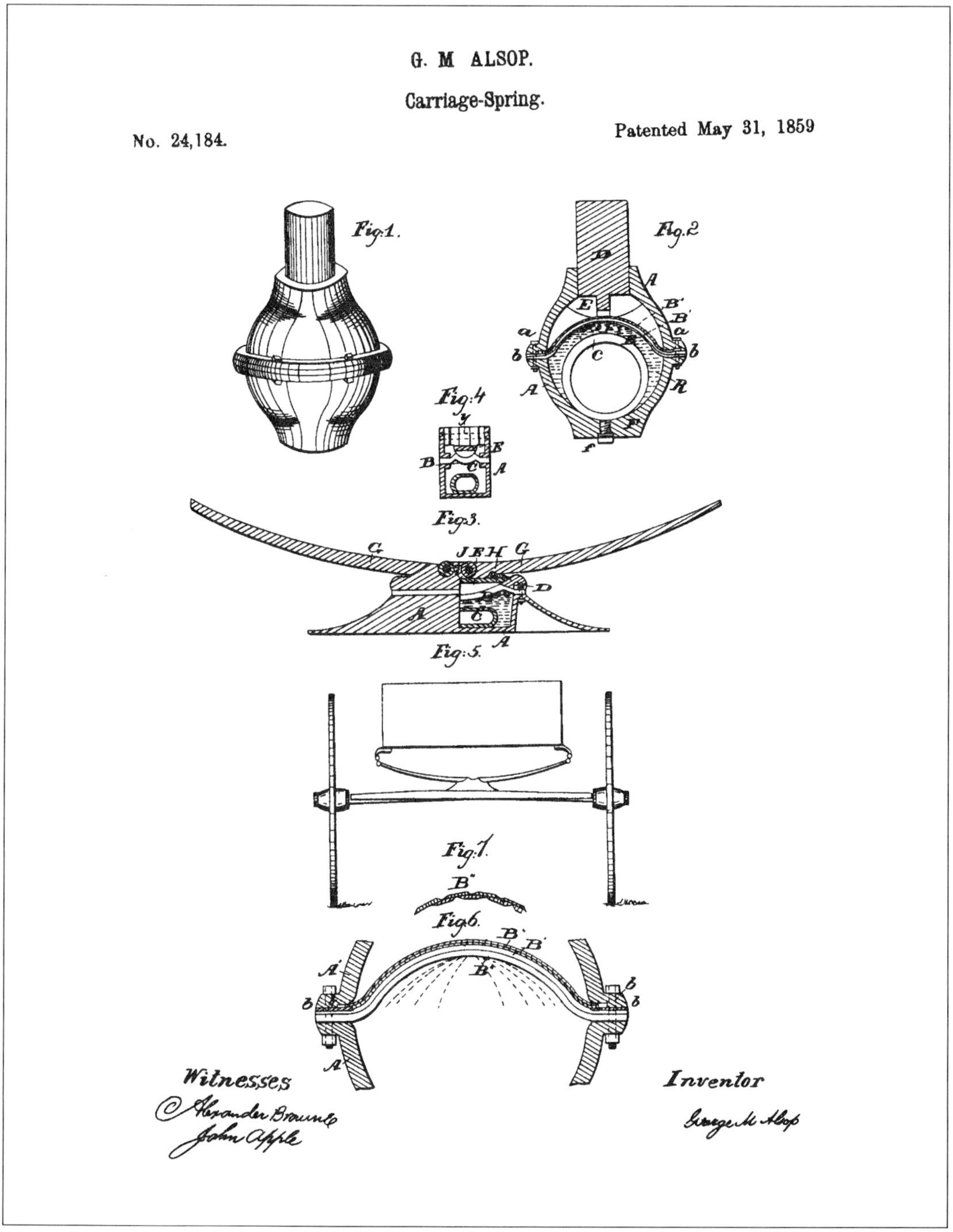

Fig. 1.4 Below the flexible diaphragm of inventor George Alsop's 1859 carriage spring, the axle of the rail car was surrounded by water.

supporting the vehicle (yet another rubberized fabric diaphragm which rolled up and down between two concentric metal cylinders) was a tiny air spring of similar design—a self-contained, jounce-actuated *air compressor* which kept the unit pumped up.

Later inventors continued to focus on the use of liquids to prevent the escape of air. In their "Improvement in Pneumatic Springs," in 1871, John Bevan and Benjamin Hitchcock of New York state devised a thick-walled two-convolution "India-rubber shell" filled with "a fluid incapable of freezing" (they suggest a mixture of glycerine and water), containing a hollow rubber ball. When the spring was deflected, the incompressible fluid squashed the rubber ball, compressing air into a metal cylinder to which it was attached (Fig. 1.5).[7]

The multiplicity of precautions against leakage, graphically documented by early designers, reflects their (justifiable) doubt about the quality of the raw materials available at the time. Long before the development of multi-roll rubber calenders which uniformly "friction and skim" high-tensile-strength nylon tire cord today, cotton fabric was rubberized (to make raincoats) by spreading a solution of rubber cement on the cloth. The coating was less than uniform, and it was often inadequately cured, because Charles Goodyear was still experimenting with his "vulcanization" recipe.

Rejecting the unreliable technology of the infant rubber industry at that time, some air spring inventors conservatively chose a sliding metal piston in a cylinder. In his 1867 patent,[8] following the lead of Lewis and Fee, W. A. Dripps further hedges against leakage of his assembly by providing a liquid interface with his sliding seal. Dripps's design (Fig. 1.6) is actually an early hydropneumatic spring, in which liquid is displaced into an adjacent chamber, where it rises to compress air above the liquid surface as the suspension goes into compression. To adjust the air pressure to tailor the spring to the supported load, liquid in the displacement chamber could be pumped in or bled out through a petcock at the bottom of the chamber.

Prototypes of these early air springs may have been produced to fulfill U.S. Patent Office requirements of the time, but there is no evidence in the literature that any of them saw the light of day in a commercial application.

The bicycle age of the 1880s and 1890s produced a complex pneumatic bicycle *seat* with "hollow cushions divided into communicating compartments [Fig. 1.7][9] capable of being inflated." The cushions were stacked on top of each other so that compression of the top one displaced air into the lower chamber.

Richard Aronstein of Goldfield, Colorado, designed a set of pneumatic bicycle springs (in combination with encircling coil springs) under both the seat and the handlebars, with provision for displacing the air into two "air reservoirs"—the interiors of tubular members of the bicycle frame. The language in Aronstein's patent reflects the need for a solution to the endless puncture problems familiar to bicyclists at the time (a vexation that has yet

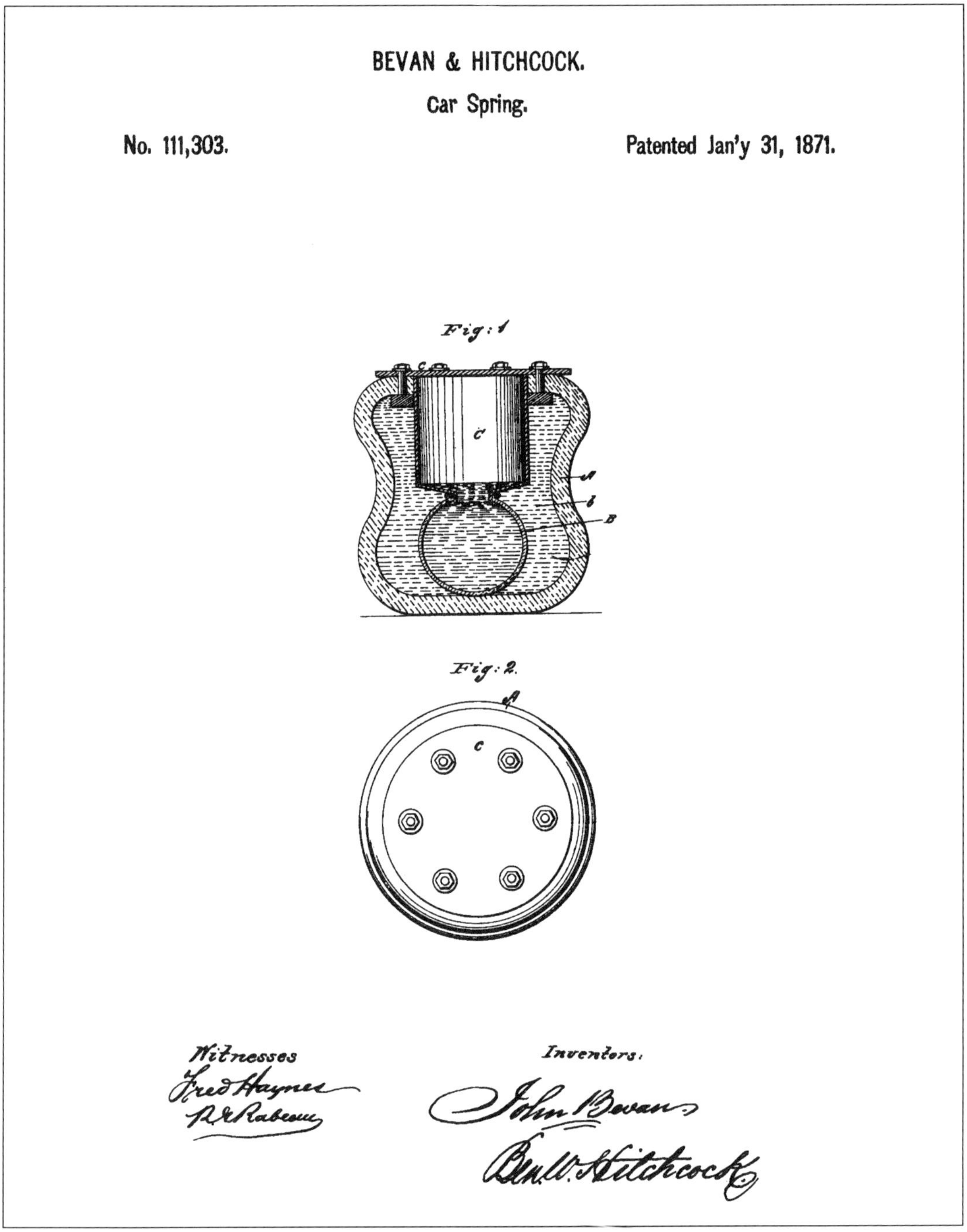

Fig. 1.5 This 1871 hydropneumatic spring designed by John Bevan and Benjamin Hitchcock employed a thick, two-convolution rubber envelope filled with a mixture of glycerine and water (to prevent freezing). When the spring was deflected, the fluid squashed an air-filled rubber ball, compressing the air into a metal cylinder to which it was attached.

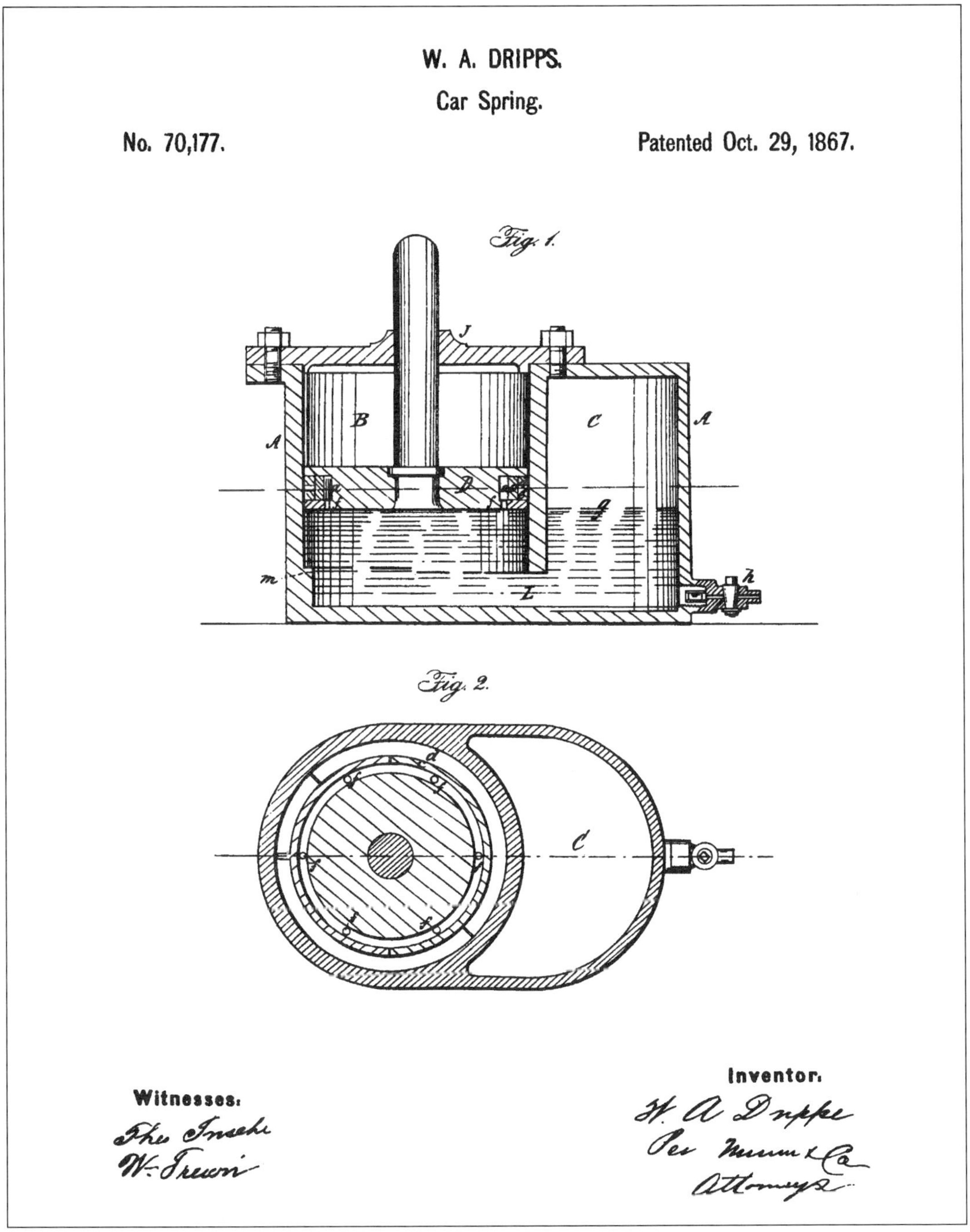

Fig. 1.6 Rejecting leak-prone rubber diaphragms of the time, W. A. Dripps resorted to a metal piston with a sliding seal for his hydropneumatic spring. When the spring was compressed into the liquid-filled cylinder, the liquid was displaced into an adjacent chamber, where it rose to compress a column of air.

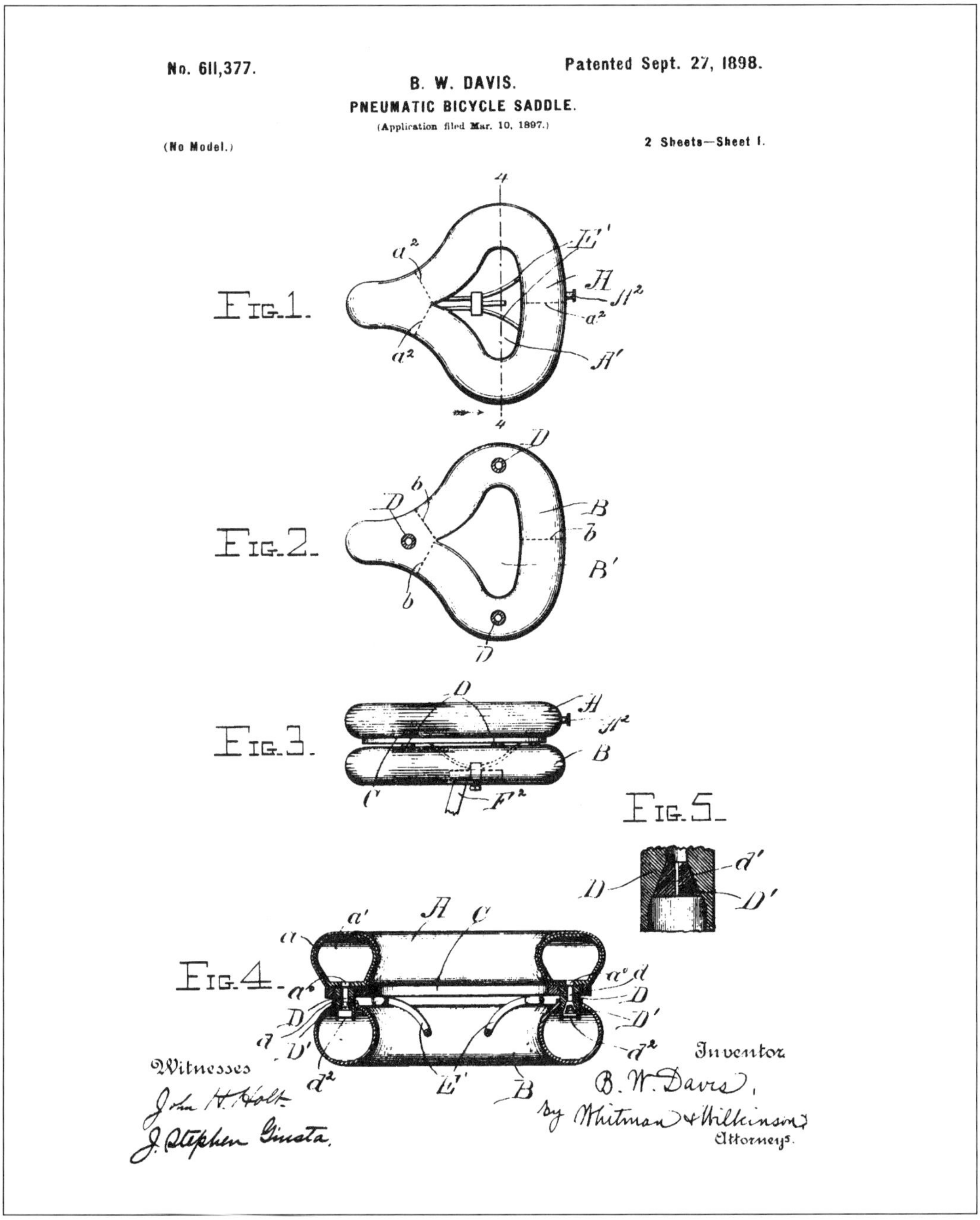

Fig. 1.7 During the bicycle age of the 1890s, B. W. Davis designed a "pneumatic bicycle saddle" which had "hollow cushions divided into communicating compartments capable of being inflated." The cushions were stacked on top of each other so that compression of the top one displaced air into the lower reservoir chamber.

to disappear a century later—particularly in California, and in other parts of the country where thorns abound). Aronstein explains that his invention has the "advantage" of permitting the use of *solid* tires, yet "giving the rider the full effect of the pneumatic tire without the many drawbacks of the latter" (Fig. 1.8).[10]

The infant automobile industry was plagued with similar problems at the turn of the century, when motor cars and trucks shared the roads with horse-drawn vehicles which shed horseshoe nails in their wake. With the life of their pneumatic tires often measured in hundreds of miles, frustrated drivers carried stacks of extra spares on cross-country trips. In an effort to provide the driver with a somewhat improved ride, trucks had pneumatic tires on the front wheels only (if at all), opting for solid rubber tires on the load-carrying rear wheels. This limited truck speeds to 15 miles per hour to prevent overheating and blowing out the melted interiors of the solid rubber tires.

In writing his 1910 patent (applied for in 1908), George Bancroft of Denver did not mince words: "It is well known that the liability to puncture problems makes the pneumatic tire objectionable on vehicles." He goes on to describe his "pneumatic element so located that it cannot be punctured, since it is supported ... entirely above ground, and it is impossible to puncture it."

Bancroft's invention, pictured in Figs. 1.9A and 1.9B,[11] consists of a hollow rubber doughnut having the appearance of a tire inner tube trapped between interlocking cones—a simple air spring interposed in series with the leaf spring suspension of the vehicle. There appears to be no provision for inflating the rubber tube.

The inventor goes on to explain that this "circular shape ... arrests shocks in all directions, whether laterally or vertically applied." An analysis of the design suggests, however, that like some other inventors Bancroft may have been more enchanted with the aura surrounding air as an isolating medium than with a means to (slightly) reduce the spring rate of the vehicle.

Unremitting problems with punctures in pneumatic tires of the time gave rise to a number of wheel-within-a-wheel inventions which employed pneumatic tires in ways that prevented the tire from ever touching the road. Alan Sullivan of Toronto conceived a pneumatic tire that was essentially a toroidal air spring, mounted *inside* the spokes of the wheel—a pneumatic doughnut sandwiched between an inner rim and the axle.[12] Several other inventors offered variations on this theme, each having a solid tire on the outer circumference of the wheel, but interposing a toroidal pneumatic cushion between an inner "hub" and the axle (Fig. 1.10).[13]

F. R. Owens's 1907 "Running Gear for Vehicles" was much more complicated. Inboard of his large-diameter road wheel (whose circumference was a flat steel rim), integral with the solid axle, but of somewhat larger diameter, was a metal roller which looked somewhat

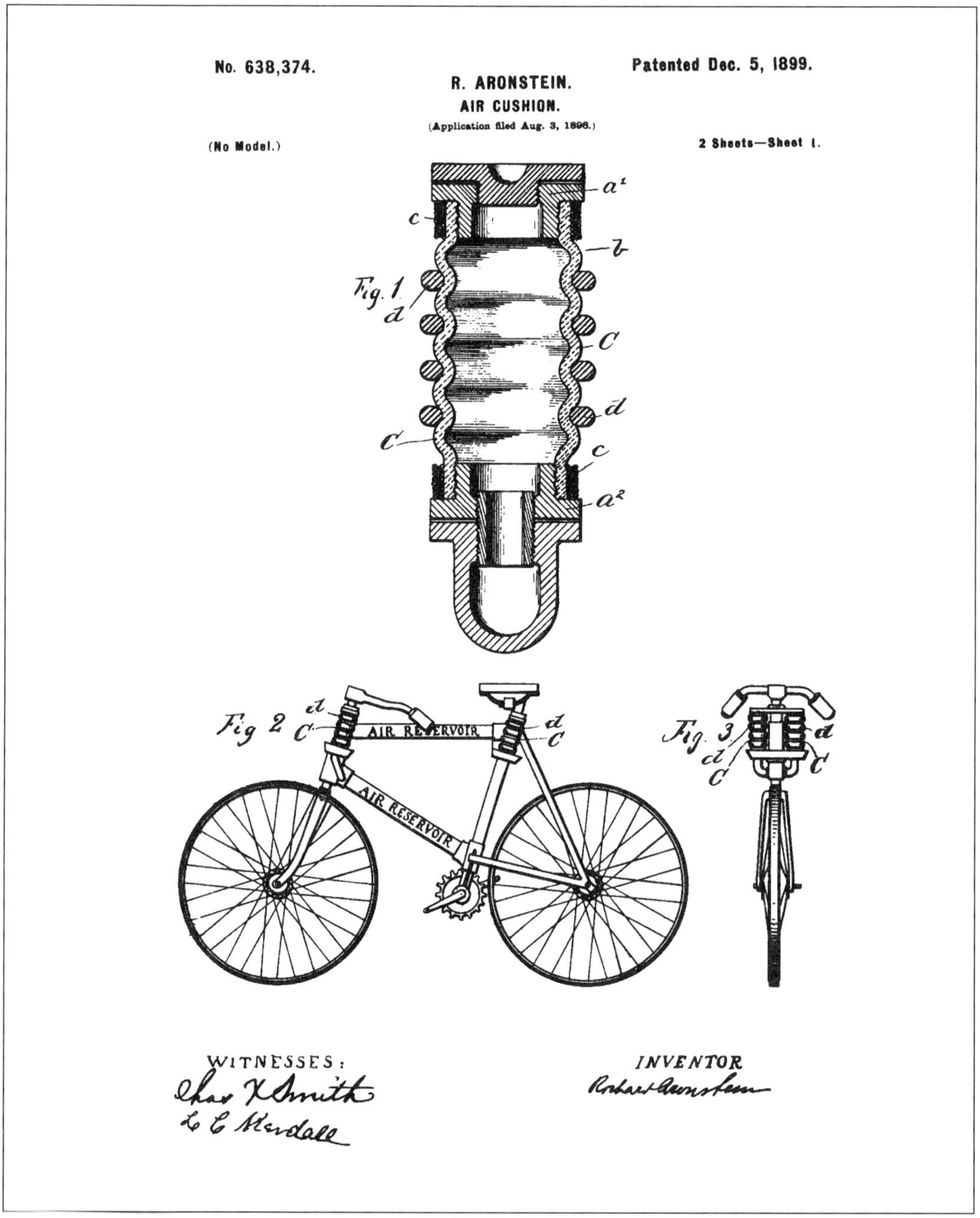

Fig. 1.8 In 1899, Richard Aronstein designed a sophisticated air suspension system for bicycles as an offbeat solution to the endless tire puncture problems encountered by bicyclists of the time. Using tubular members of the bicycle structure as displacement reservoirs, his air spring (encapsulated by a coil spring) had the "advantage," he explained, of permitting the use of ***solid*** *tires, yet "giving the rider the full effect of the pneumatic tire without the many drawbacks of the latter."*

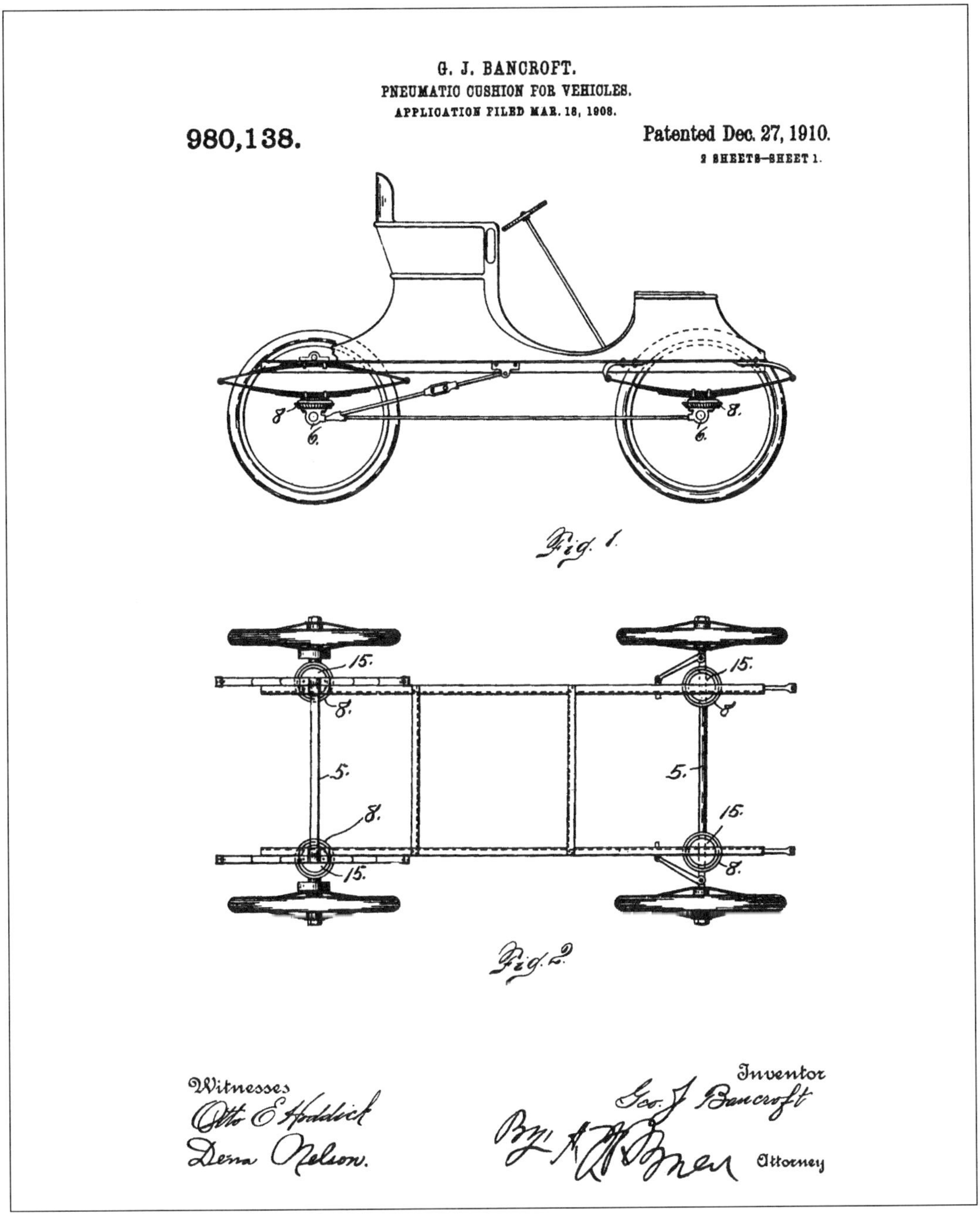

Fig. 1.9A With the growing popularity of motor vehicles, George Bancroft's 1910 invention had a goal similar to Aronstein's: "It is well known that the liability to puncture problems makes the pneumatic tire objectionable on vehicles." He explains that his "pneumatic element [is] so located that it cannot be punctured, since it is supported ... entirely above ground, and it is impossible to puncture it."

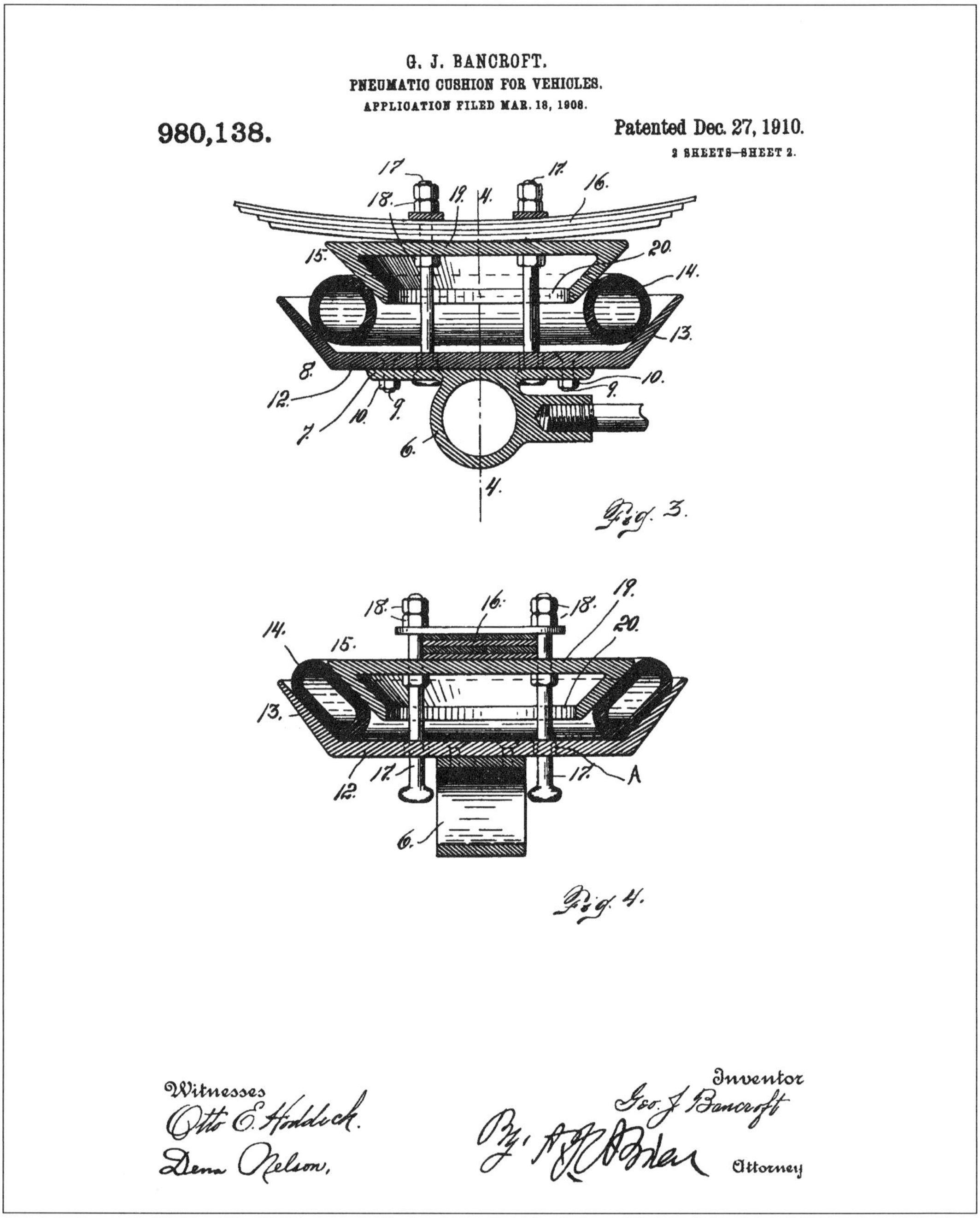

Fig. 1.9B Bancroft's air spring was a hollow rubber doughnut in series with the leaf spring of the car.

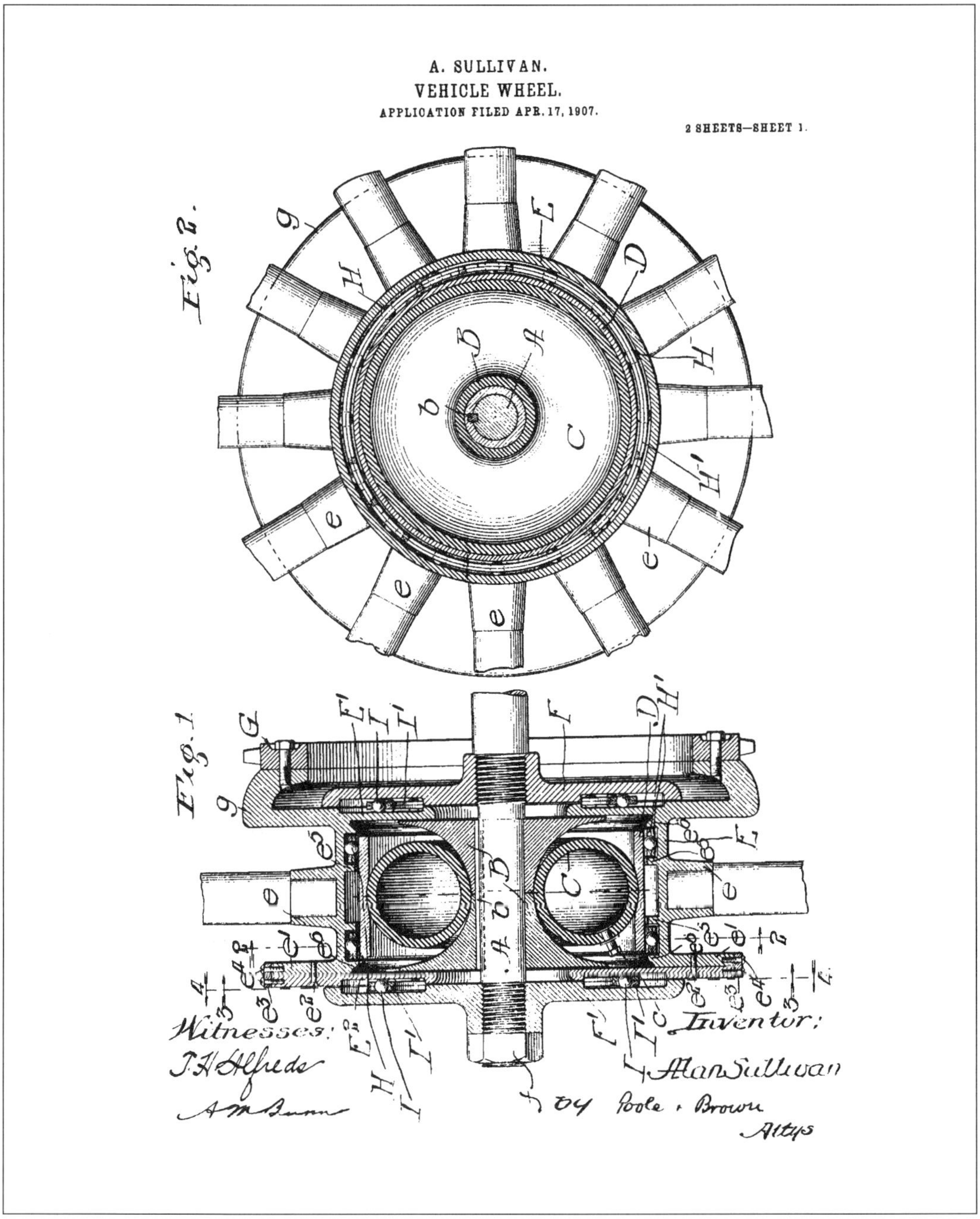

Fig. 1.10 Unremitting puncture problems on passenger cars inspired several "wheel-within-a-wheel" inventions which sought to retain the pneumatic tire, but to prevent it from touching the ground. Alan Sullivan's 1908 device was essentially a toroidal air spring, mounted ***inside*** *the spokes of the wheel—a pneumatic doughnut sandwiched between an inner rim and the axle.*

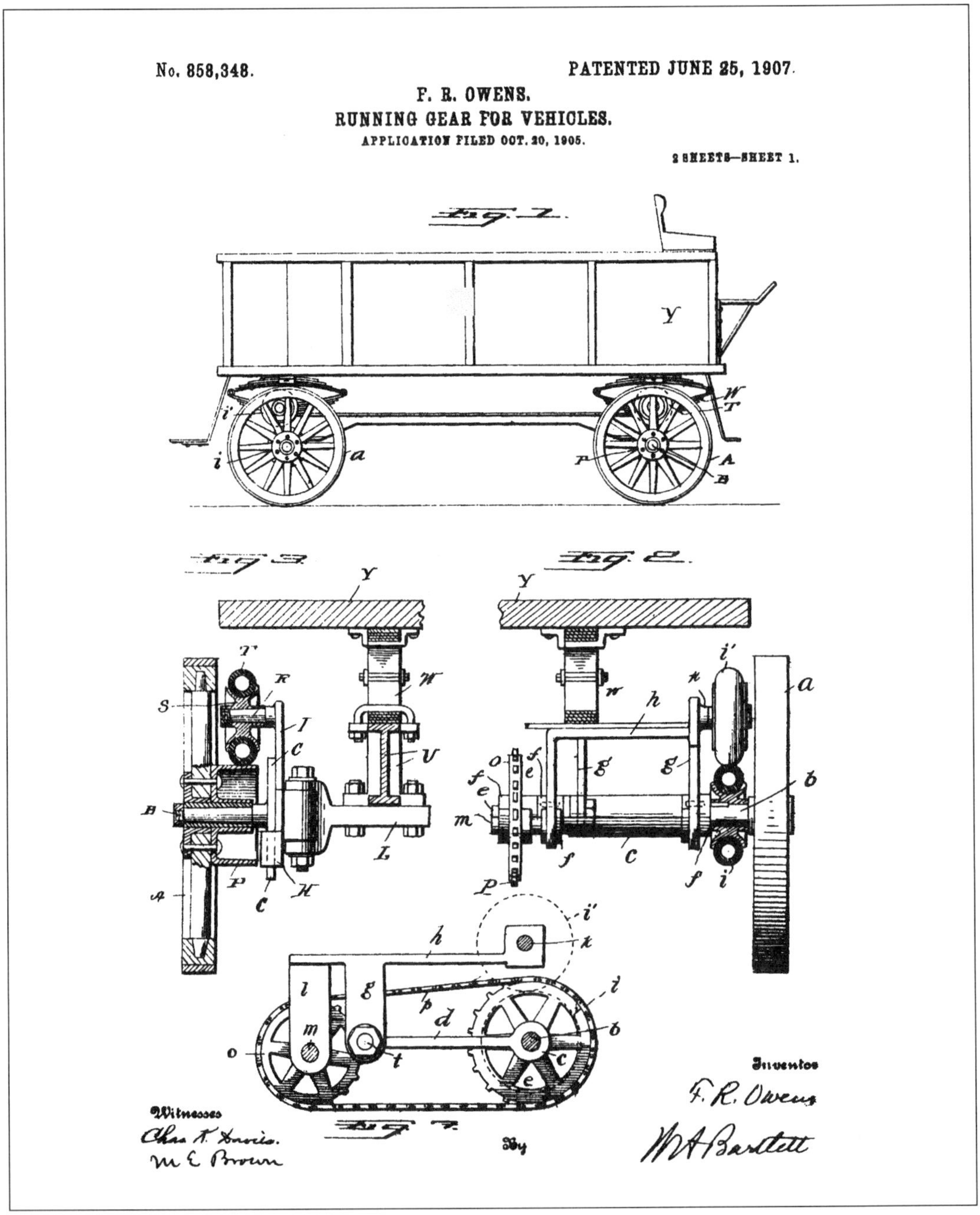

Fig. 1.11 F. R. Owens's 1907 "Running Gear for Vehicles" was more complicated. Inboard of his large-diameter road wheel was a metal roller which looked something like a brake drum. Resting on the roller, and running on its machined cylindrical surface instead of on the ground, was a smooth-tread pneumatic tire, which became a rotating air spring. Once again, the pneumatic tire was prevented from touching the road—which, at the time, was often littered with horseshoe nails.

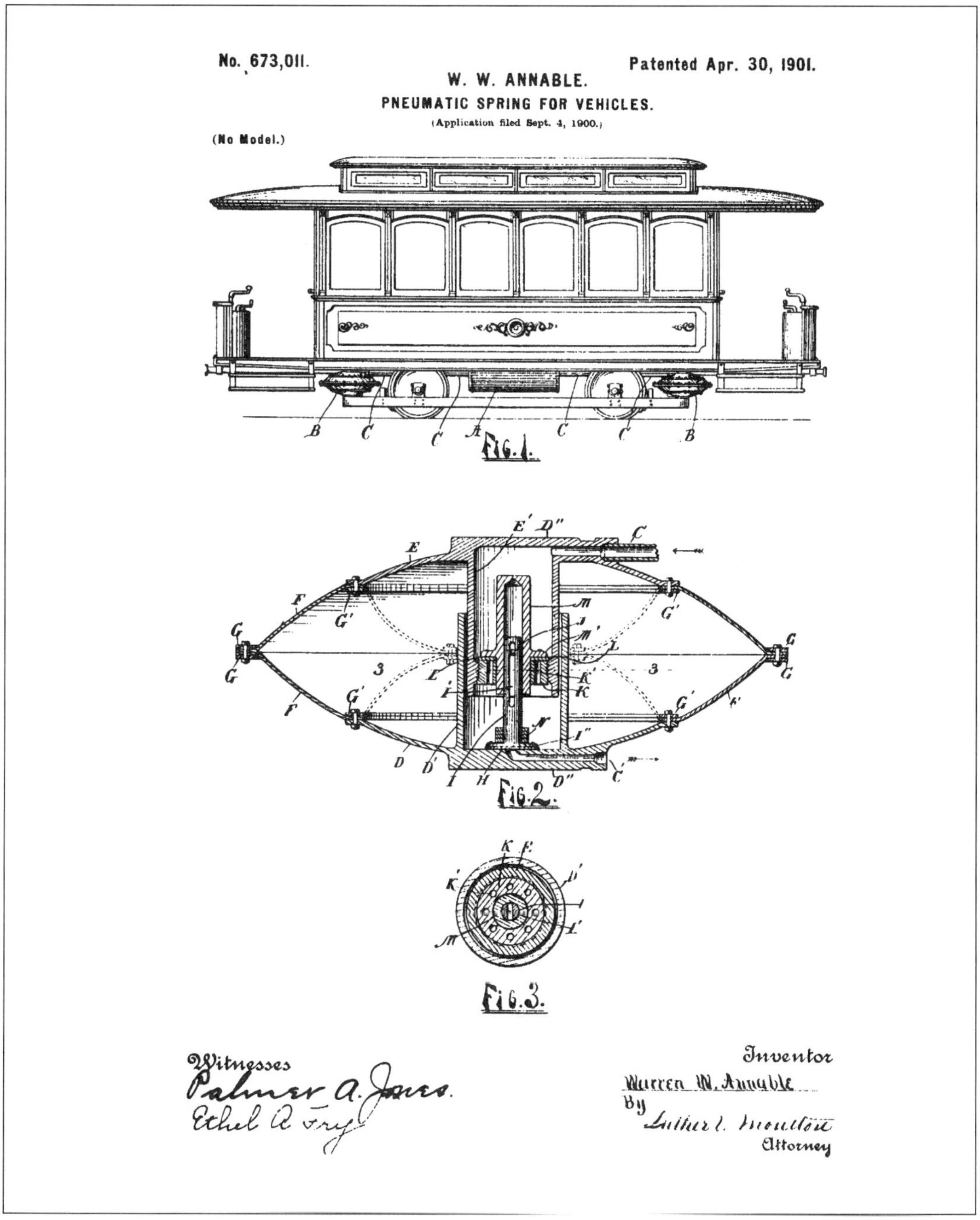

Fig. 1.12 Perhaps the first patent for air springs on a transit system vehicle, Warren Annable's 1901 patent was an air spring for street cars. But it would be half a century before pneumatic suspensions would become a reality—on transit ***buses****.*

like a brake drum. Resting on the roller, and running on its machined cylindrical surface instead of on the ground, was a smooth-tread pneumatic tire, which thus became a rotating air spring (which is, of course, what all pneumatic tires are). And, in case this arrangement didn't provide sufficient cushioning, Owens mounted a stacked elliptical leaf spring above and in series with his protected pneumatic tire (Fig. 1.11).[14]

As early as 1901 an air spring patent was granted for electric street car suspensions (Fig. 1.12).[15] But it would be on diesel-powered buses half a century later that the first commercial air spring installations would appear. However, they wouldn't achieve success until post-World War II polymers, including nylon tire cord and synthetic rubber elastomers (especially Neoprene) came along.

Chapter Two

REDUCTION TO PRACTICE

One of the first production motor vehicles in the United States to be equipped with air springs was the 1914 Owen Magnetic, shown in Fig. 2.1. The unusual name of this car referred to its drive system. Although the Owen had a conventional (initially 38-horsepower) gasoline engine, its engine drove a generator instead of a geared transmission. The resulting electric current fed an electric motor which turned the wheels—a system not unlike that employed on diesel-electric trains decades later. The Owen's metal piston/cylinder air springs were precursors of hydropneumatic suspensions that would appear in Europe nearly half a century later (see Chapter 9).

There was no bumper in front of the Owen's radiator; instead, the car's styling was dominated by two black and gold vertical pneumatic cylinders with silver air fittings at the top. Each of these assemblies was attached between the front end of a longitudinal leaf spring and the frame, serving as the spring shackle. Their lower ends full of oil, the cylinders contained sliding pistons sealed by sliding leather rings, with an air chamber in the top half. These hydropneumatic springs were, thus, in series with the steel spring—like some trailer suspensions half a century later. A similar arrangement was provided on the rear suspension, as seen in Fig. 2.2.

This air-leaf suspension was one of 400 inventions patented by innovator/entrepreneur George Westinghouse (1846–1914), inventor of the railroad air brake and founder of some 60 companies—including the Westinghouse Electric Company, pioneer of alternating current. Made in Cleveland, these Westinghouse springs also appeared on the 1912 Packard, the 1913 Simplex, and on some early Hupmobile and Peerless models.

The early twentieth century also produced an assortment of flexible (fabric-reinforced-rubber) air spring designs which departed substantially from prior art. Most led nowhere. Examples include Alfred Weiland's 1926 patent for a pneumatic "shock absorber" (Fig. 2.3),[1] whose accordion-like bellows linkage is similar to that of classic blacksmith shop bellows. The durability of such a configuration in a vehicle suspension, however, is questionable to say the least. J. A. Heany's 1928 concept (Fig. 2.4)[2] consisted of a *pair* of rubber bladders—one for the jounce stroke and one for rebound.

Fig. 2.1 A pair of Westinghouse hydro-pneumatic springs dominated the front end of the 1916 Owen Magnetic, one of the first motor vehicles equipped with air suspension. The name of the car related to its drive system—a gasoline engine turning a generator whose current drove an electric motor. Serving as the spring shackle, the piston/cylinder assembly was attached between the front end of a longitudinal leaf spring and the frame. Projections on the top are air valves through which the springs were manually pressurized. (Courtesy of the Crawford Museum, Western Reserve Historical Society. Photo by the author.)

Fig. 2.2 Rear end of the Owen Magnetic, showing how the Westinghouse hydropneumatic pistons were attached to the ends of the rear leaf springs. (Courtesy of the Crawford Museum, Western Reserve Historical Society. Photo by the author.)

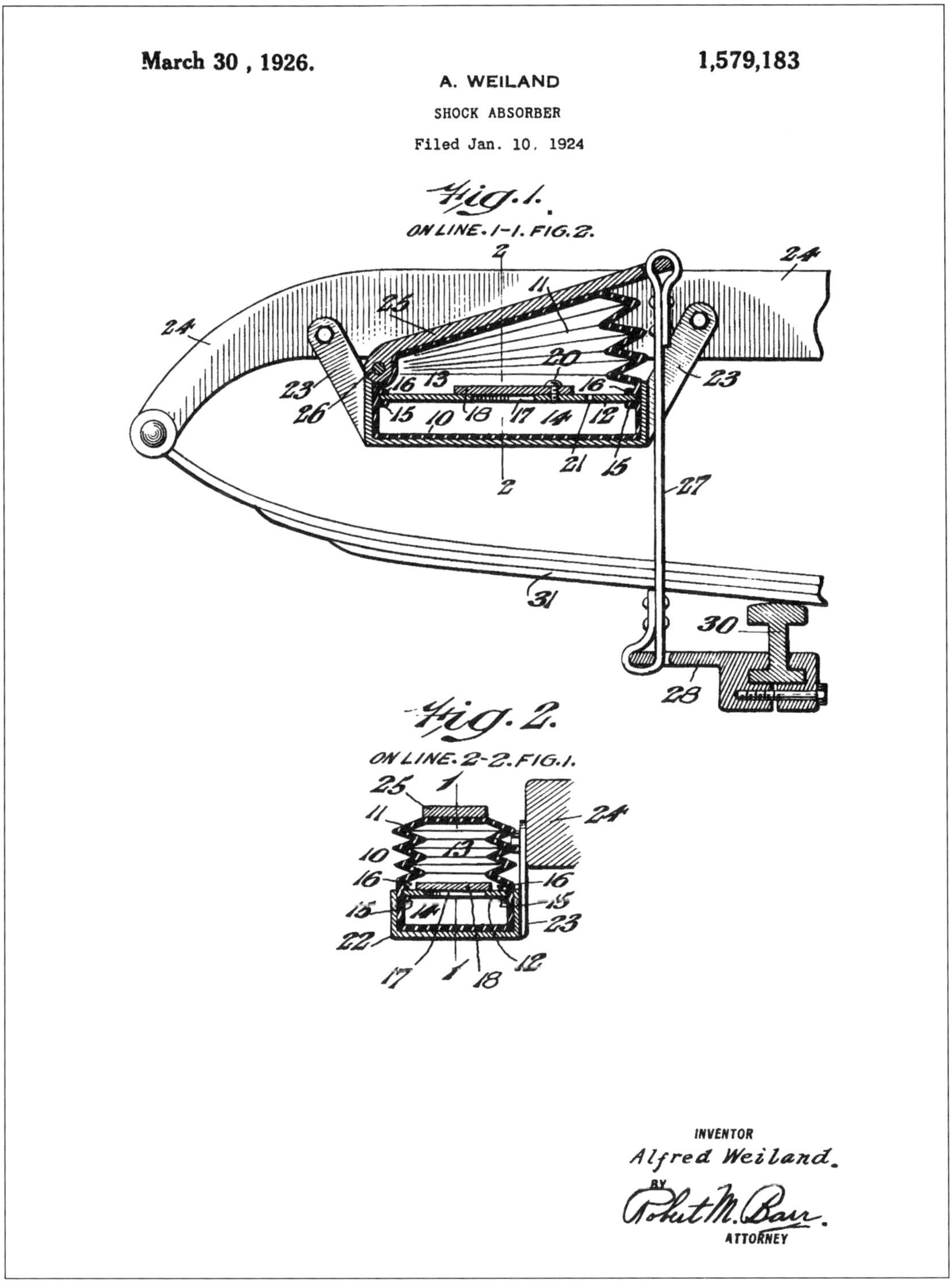

Fig. 2.3 Alfred Weiland's 1926 patent—an accordion-like air spring configuration that resembled a blacksmith's bellows.

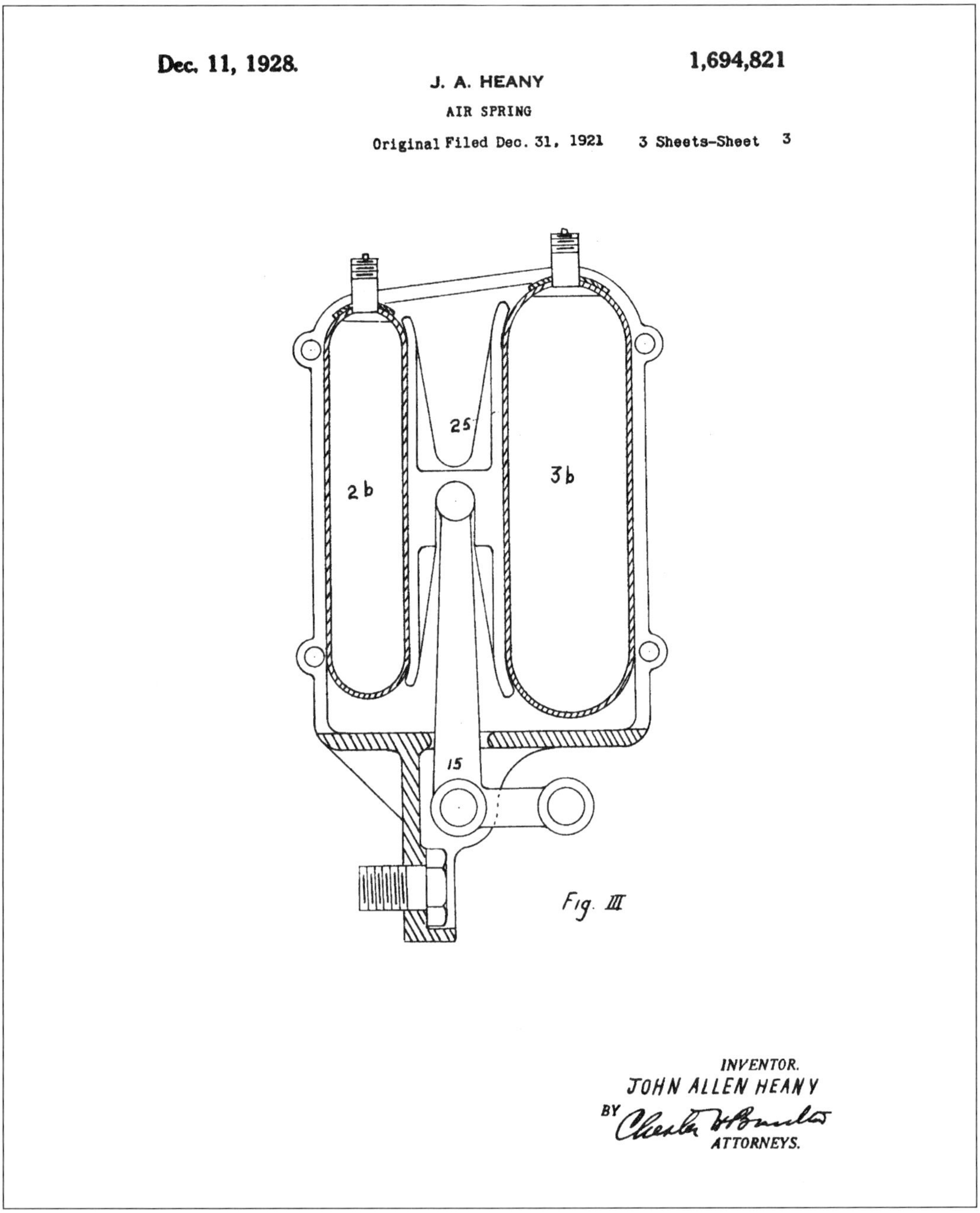

Fig. 2.4 This bell-crank air suspension geometry by inventor John Allen Heaney uses two inflated balloon-like bladders—one for jounce, the other for rebound.

The inflated cushions are alternately compressed by the rocking motion of a crank mechanism attached to the axle. Although his design would work, the resulting ride would be disappointing.

But 15 years *earlier*, between 1910 and 1914, three U.S. patents anticipated sophisticated concepts which were subsequently incorporated into pneumatic suspension systems that went into production decades later. The first of these was issued to Benjamin Bell, "a subject of the King of Great Britain, residing in Philadelphia." It is pictured in one embodiment in Figure 2.5.[3] With its guided reversible diaphragm whose lower end is fixed to a piston and a formed metal air chamber into which the air is compressed as the piston moves upward in the jounce stroke, Bell's pneumatic spring bears a remarkable resemblance to the Type A[4] GM Technical Center passenger car air spring discussed in Chapter 5, as incorporated on the 1958 Buick and Oldsmobile.

Having experimented with a variety of piston shapes at the time, Bell discloses, perhaps for the first time, the concept of controlling the *effective area* characteristic of a pneumatic spring by means of a contoured concentric metal skirt surrounding the flexible diaphragm. Disclosing his innovation, Bell's patent states, "The angle of the flaring portion **12** may be changed so that the area of pressure acting against the members **3** and **4** will increase with movement of the members **3** and **4** away from each other, and decrease the movement toward each other, in any desired manner, either to maintain a substantially constant resultant pressure against members **3** and **4** ... irrespective of their relative positions, or so that the resultant pressure against said members may vary when they come to different relative positions, to suit any desired requirement of work of the pneumatic spring."

Explaining the design flexibility afforded by the outer metal guide, Bell's patent teaches that, "by its use and variation, any desired load may be carried by the upper member, supported by the lower member ... the angle of the flaring portion determining the resultant pressure against the members **3** and **4** ... compensating by increased air pressure area for the decreased air pressure per square inch as the members **3** and **4** are moved in a direction away from each other."[3]

An analysis of the changes in effective area that occur as this spring operates (with reference to a much longer explanation in Chapter 4) will reveal that as the piston moves upward in jounce stroke, the center of the curved meniscus of the diaphragm moves *inward* toward the center of the spring, incrementally decreasing the effective area, *and* the slope of the resulting load-deflection curve—thus softening the spring.

Two other very significant patents at the time were issued to Joseph Hofmann, "a subject of the Emperor of Austria-Hungary, residing at Baumaroche, Switzerland." Like Bell's, Hofmann's innovations (see Figs. 2.6 and 2.7),[5] include an air spring consisting of a flexible diaphragm whose contour is also guided by formed metal stampings which

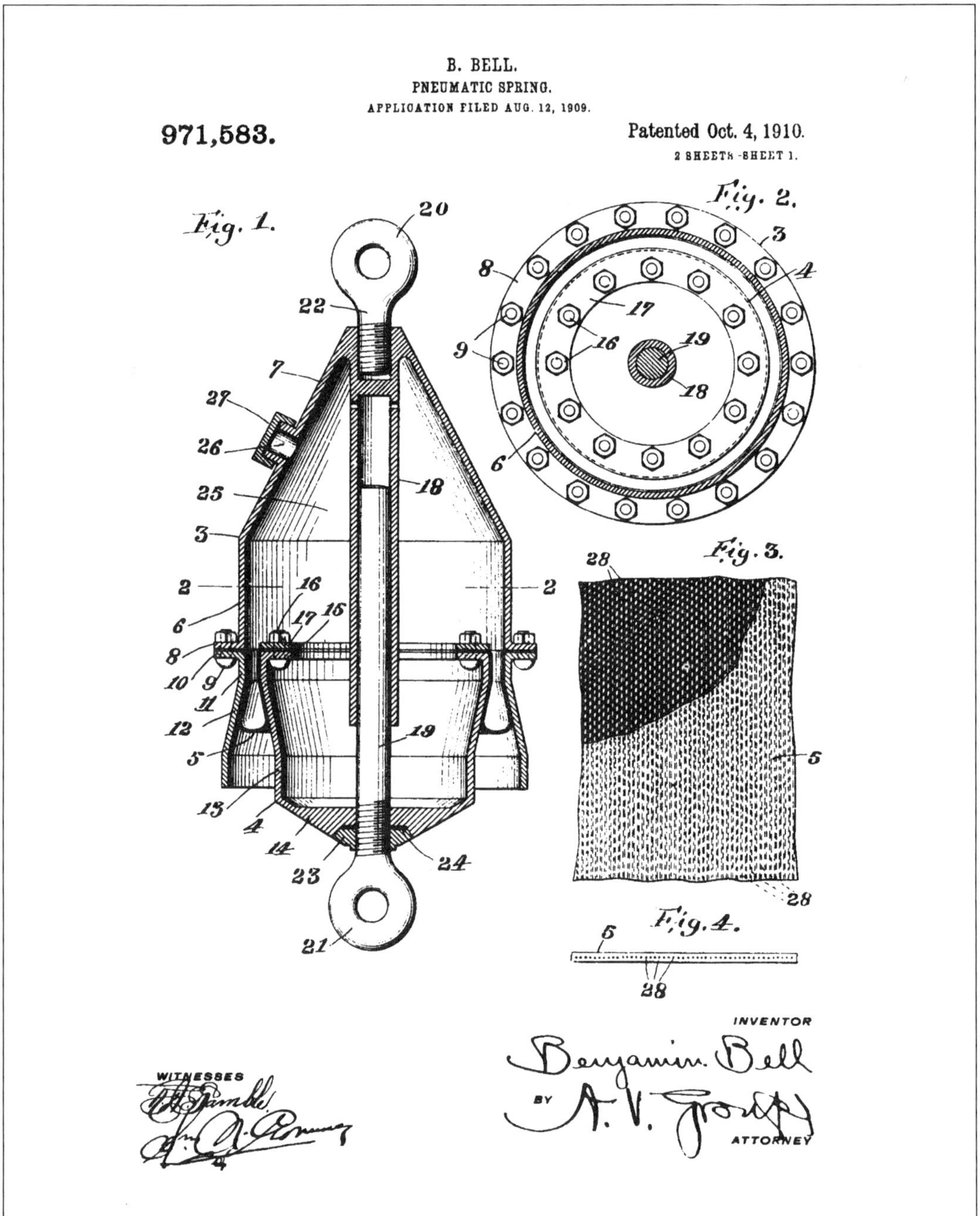

Fig. 2.5 Almost half a century ahead of its time, Benjamin Bell's 1910 pneumatic spring design may be the first to utilize a contoured piston and shaped metal skirt to control changes in the effective area of an air spring to tailor its load-deflection curve—a concept employed in the 1950s by the General Motors Technical Center air spring design released on 1958 Buicks and Oldsmobiles.

control the effective area of the spring throughout its vertical stroke. But the Hofmann patent also describes and illustrates a sophisticated *valve* in a pipe between the air spring and an air chamber into which air is displaced by the pneumatic spring. The valve is "operable by gravity and centrifugal force, and capable of closing said pipes to suddenly diminish the effective volume of the air cushions." One sees Hofmann's influence not only in Roy Brown's prototype passenger cars of the 1930s, described below, and in Brown's 1936 patent, but also in some of the air suspension passenger cars of the 1950s.

By 1934, two years before Brown's patent issued, A. E. Forsyth had disclosed a pneumatic suspension design which floated a passenger car body on two-convolution flexible rubber air springs—one at each wheel (Fig. 2.8).[6] Engineers had obviously begun to give some thought to air suspension *systems* for passenger cars.

Roy Wilbur Brown (1891–1967), a Firestone engineer who managed a small aircraft product development group called "Research Engineering," located in the company's Akron, Ohio, headquarters during the 1930s, was the father of the two-convolution "bellows" configuration that became the first successful air spring. Although fabric-reinforced rubber bellows were not new, it was Brown who devised a manufacturing method that (with refinement by others) eventually led to a durable product which would be capable of operating literally for decades in the harsh environment of a vehicle undercarriage.

Brown's two-convolution reinforced rubber bellows looked like a stacked pair of small tires lying on their sides, joined at the beads—and, from a construction perspective, this is essentially what they were. The reinforcement, which permitted the bellows to contain internal air pressure without ballooning, was tire-cord fabric—a special weave in which the parallel load-carrying cords (warp) are held together with just enough weak "pick" cords (fill) to provide stability to the fabric during the tire-building process. To achieve adhesion to rubber, tire cord is dipped in a bonding cement before the fabric is "frictioned and skimmed" with rubber (or, today, synthetic elastomers) by the steel rolls of a "calender."

Brown's bellows were built like a tire on a simple cylindrical drum that looked like a miniature tire-building machine. Layers of unvulcanized rubber and rubber-coated tire fabric were rolled on and "stitched" down to adhere the tacky layers together before blowing them off the drum with an air nozzle.

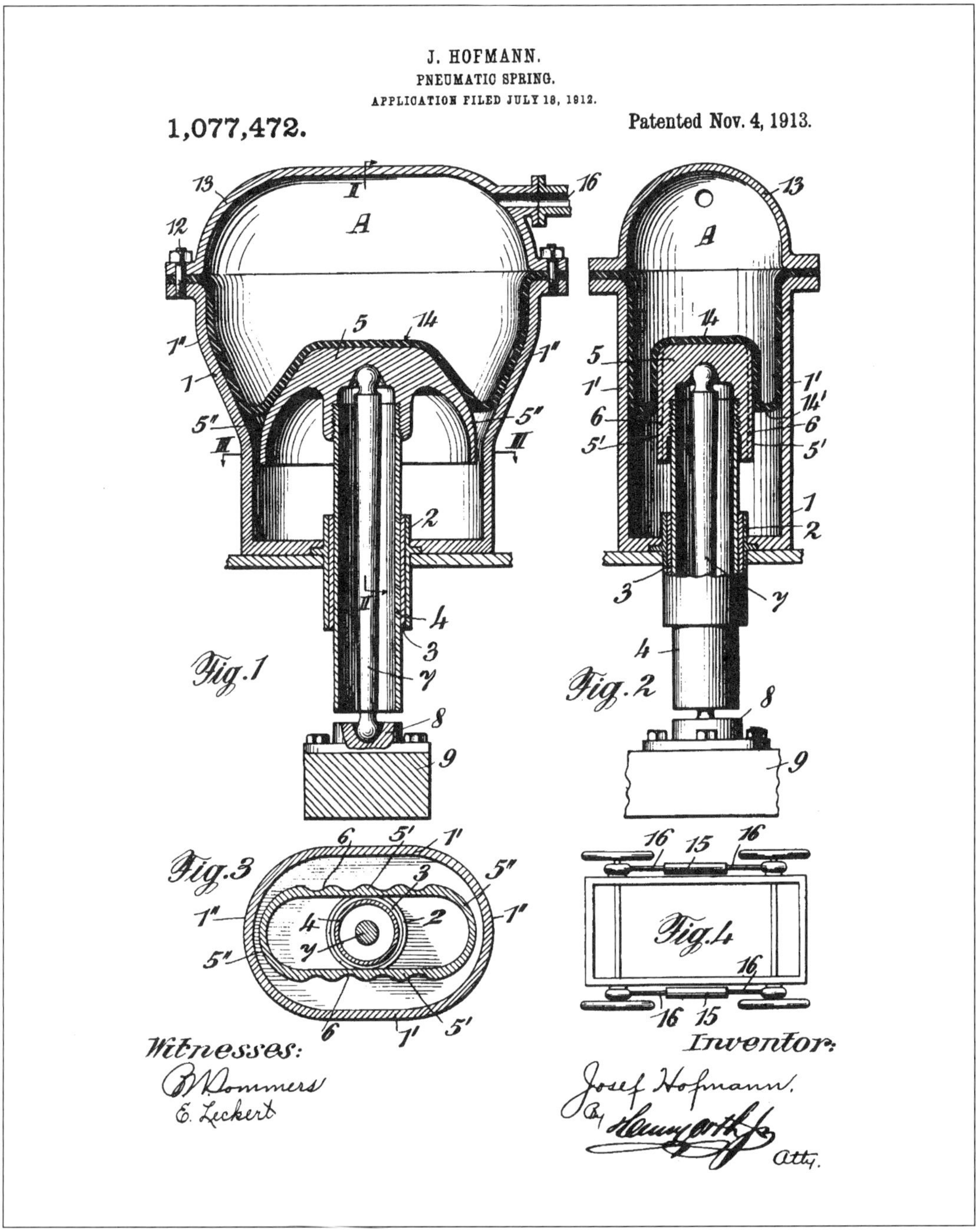

Fig. 2.6 Joseph Hofmann's air spring also used contoured metal components to control the effective area of its flexible diaphragm. But Hofmann's design had other innovations that are revealed in the figure following.

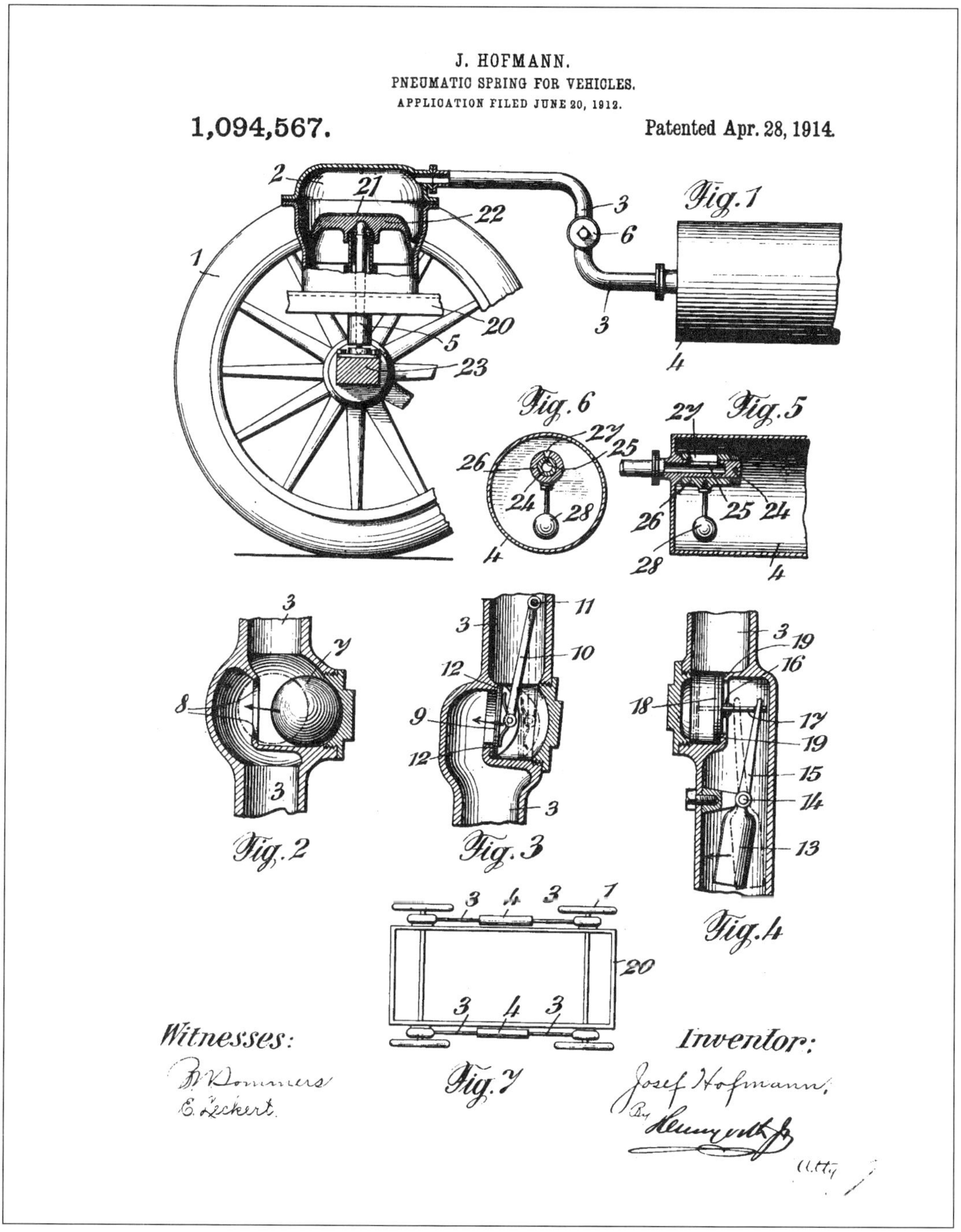

Fig. 2.7 Hofmann's suspension featured a valve, "operable by gravity and centrifugal force, and capable of closing said pipes [between the air spring and its displacement reservoir] to suddenly diminish the effective volume of the air cushions"—an early attempt at an "active suspension."

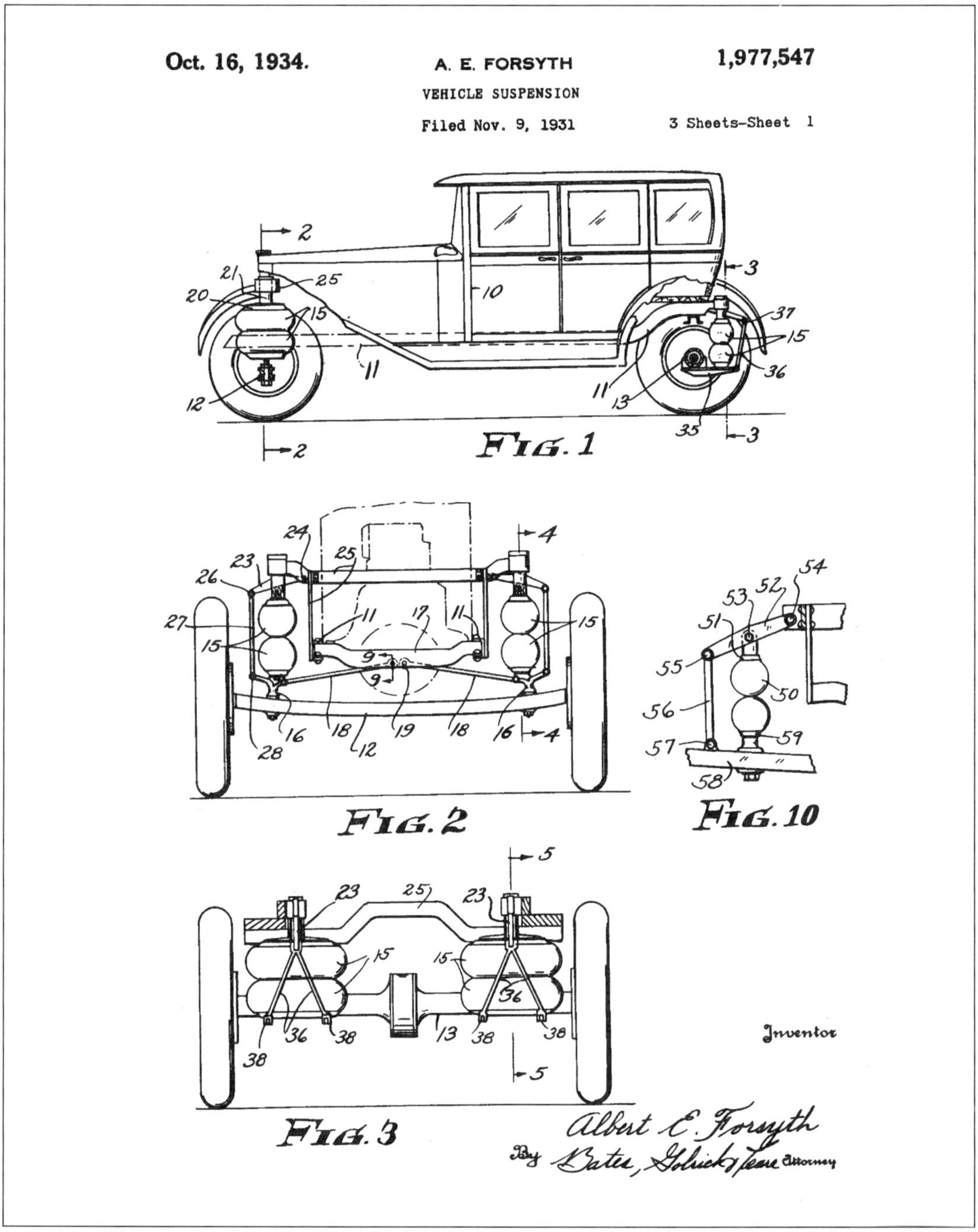

Fig. 2.8 One of the first "full-air" pneumatic suspension designs for passenger cars, Albert Forsythe's 1934 patent featured a two-convolution air spring at each wheel.

Using long-established tire technology, Brown's technicians built up a cylindrical assembly consisting of several layers: To contain the air like a modern tubeless tire, a thick, unreinforced rubber "inner liner" was first applied to the building drum and its cut ends were spliced together and rolled down. This layer was followed by two plies of rubberized tire cord, with the cords intersecting at an acute angle, as seen in Fig 2.9. On top of these plies, an abrasion-resistant cover layer was rolled down. The cords at each end of the sandwiched rubber and fabric cylinder were then flared ("turned up") and wrapped around hoop-shaped rings made of aircraft cable to hold them in place during the molding operation. This stranded-cable "bead wire" (which, in later production, became a solid steel ring) also provided a stiff interior member within a molded rubber lip around which a sealing device could later be clamped.

The resulting barrel-shaped "mold prep," having the appearance sketched in Fig. 2.10, was expanded into a steel mold of one, two, or three convolutions, curing the part (vulcanizing the rubber) against the hot mold surface, with pressurized superheated water (later steam) on the interior. Examples of the resulting single, double, and triple convolution molded "bellows" produced by this method are shown in Figs. 2.11 and 2.12.[7]

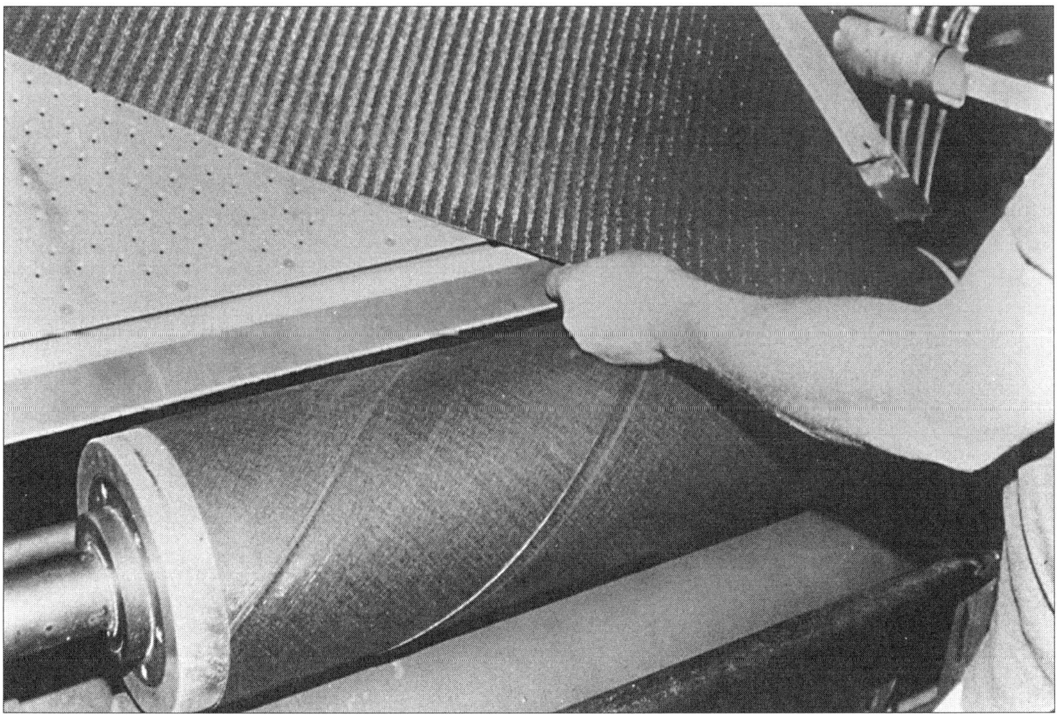

Fig. 2.9 Building a bellows-type air spring: Following a layer of rubber inner liner, two plies of "bias-cut" rubberized tire cord fabric are being wrapped on the building mandrel with their cords intersecting at an acute angle. These will become the reinforcing layers that contain the pressurized air in the molded part. (Firestone Industrial Products Co.)

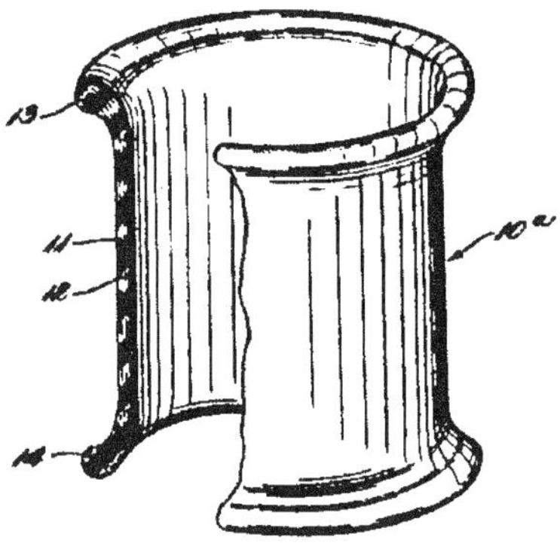

Fig. 2.10 The barrel-shaped "mold prep" for a convoluted air spring has metal beads at both ends, around which the tire cords are wrapped.

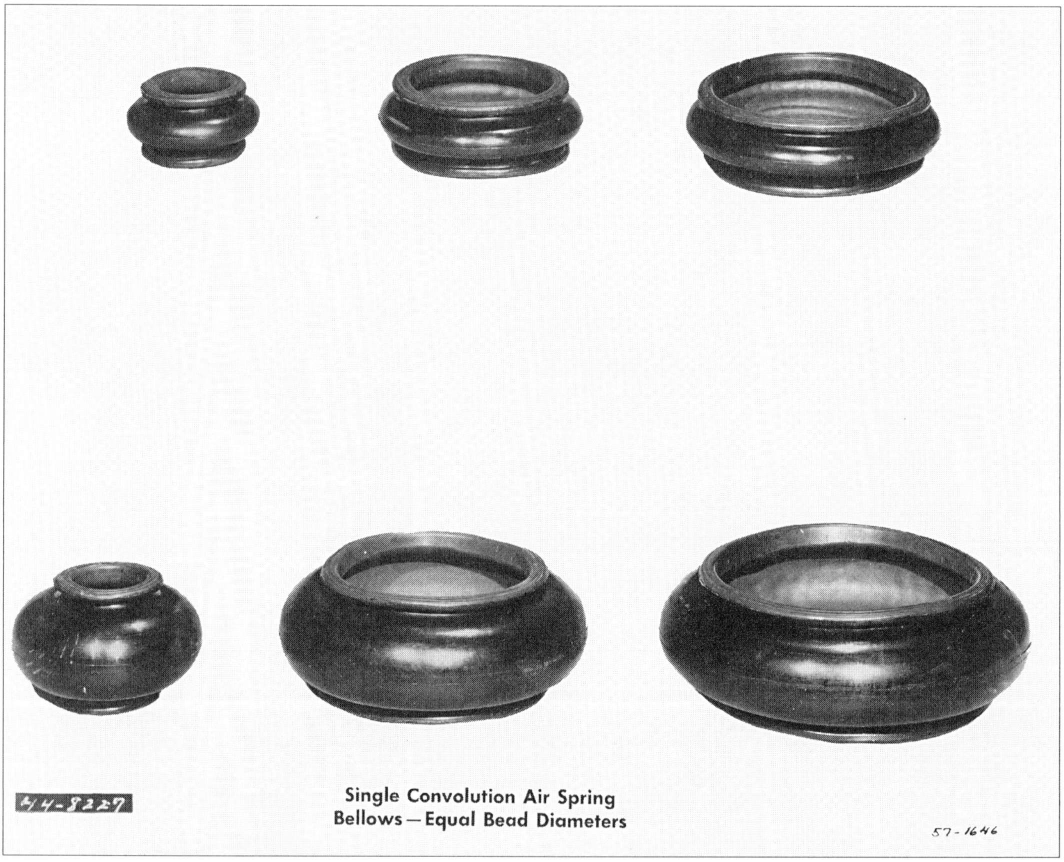

Fig. 2.11 Early single-convolution bellows-type air springs made by Firestone, from molds that date back to the 1930s. (Firestone Industrial Products Co.)

Fig. 2.12 A variety of two- and three-convolution Firestone air springs made in the 1930s and '40s. Some of these are still marketed today. (Firestone Industrial Products Co.)

Metal "girdle hoops," assembled after cure on early production, would restrain the small-diameter circumferential groove(s) between the bulging convolutions when the unit was later pressurized. When assembled into the vehicle suspension, cast bronze or, later, malleable iron (and, still later, aluminum) "bead rings" and bolts would clamp the molded sealing surfaces on both ends of the bellows to adjacent metal components, as shown in Fig. 2.13.[8]

Attracted by the potential volume of passenger cars, Brown targeted that market almost exclusively for nearly a decade. To demonstrate his product, he modified several existing car models, replacing the steel springs on a 1934 Plymouth, a 1935 Buick (Fig. 2.14), a 1936 Chrysler (Fig. 12.15), a Ford, a Cord (Fig. 2.16—in process of installation), a Studebaker, a LaSalle, a Lincoln, a Checker Cab, and an experimental rear-engine Stout

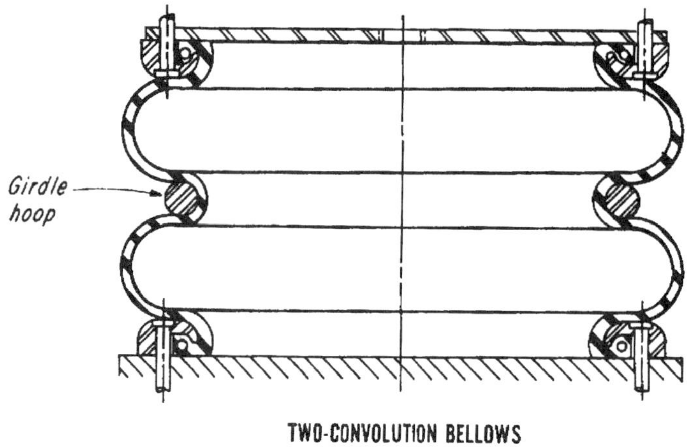

Fig. 2.13 The rubber beads on Roy Brown's early two-convolution air springs were sealed to mating surfaces with cast "bead rings" and bolts. A "girdle hoop," which prevented the middle of the bellows from ballooning when pressurized, was assembled after the part was cured.

"Scarab"—converting each to air suspension. Visible in each of these photos is (1) the air spring, (double-convolution bellows), (2) a large-diameter hose (about an inch [25 cm]), which is connected to (4) a "reservoir" tank containing somewhat more volume than the air spring itself. This additional "displacement volume" was necessary to decrease the rate of change of pressure build-up in the air spring, thus reducing its spring rate.

Fig. 2.14 Roy W. Brown's experimental 1935 Buick, which he converted to air suspension for demonstration to the Society of Automotive Engineers in 1936. (Firestone Industrial Products Co.)

Fig. 2.15 A 1936 Chrysler converted to air suspension by Brown. (Firestone Industrial Products Co.)

Fig. 2.16 Another experimental Brown vehicle—a Cord in process of conversion to air suspension. (Firestone Industrial Products Co.)

Inside the upper convolution of the bellows was a "pendulum valve" (Fig. 2.17),[9] which, when swung sideways by lateral centrifugal forces as the car rounded a curve, would momentarily close the passage to the displacement reservoir tank—effectively stiffening the outboard spring to limit roll of the vehicle (which could otherwise be excessive with the relatively low spring rate for the time). Similarly, forces in the fore-and-aft direction also caused the valve to close, limiting dive upon braking.

Brown's valve action was similar to that proposed by Hofmann two decades earlier. But Brown's valve had an additional function: As his patent explains, the valve is also actuated by the *vertical* accelerations created by the direction reversal of the suspension from the compression (jounce) stroke to the (rebound) stroke. Closure of the valve momentarily delayed return of air from the reservoir, a feature intended to limit rebound—an example of Brown's attempt at *air damping*, a lifelong ambition. His suspensions had no shock absorbers. A schematic of Brown's chassis is shown in Fig. 2.18.[9]

Roy Brown drove several of his cars to annual meetings of the Society of Automotive Engineers in Detroit, and, according to author Fred Haushalter, once set up a demonstration for Henry Ford, arranged by Harvey Firestone, Sr. Without explaining the details of his suspension, Brown drove his two famous passengers over a variety of Dearborn and Detroit-area rural roads, probably including one of his favorite surprising side-trips—driving down railroad tracks over the ties.

When their ride ended, Ford asked to have the car put up on a hoist in the Ford garage. After silently examining the complex suspension (an air bellows at each wheel with a fat rubber hose running to an air tank displacement reservoir, a smaller air line routed to an air compressor, and Brown's sophisticated pendulum valve above each air spring), Ford "emphatically told his friend Harvey that the only suspension that was any good was his transverse [leaf] spring suspension, which he had used from the beginning of his automobile career." (And which, Haushalter adds in an editorial aside, "everyone who owned a Ford car knew was very unstable in a side wind.")[10] Salesman that he was, Firestone didn't argue. And those of us who knew Roy Brown find it surprising that he didn't either.

Brown's passenger car design was both too complicated and too costly for the automotive industry to even consider in the midst of the Great Depression. Moreover, the relatively minor variation in the weight of a passenger car due to payload changes (compared to the tare weight of the car) didn't really warrant a sophisticated variable-rate suspension. Besides which, adding to the complexity and cost, Brown advocated provision for adjusting the stiffness of his air springs (either manually or by automatic means) not only to accommodate changes in vehicle payload, but "to approximate the ideal rate condition necessary for the boulevard ride, and also to provide the increase in rate desirable … when driving at high speeds or over rough roads."[11]

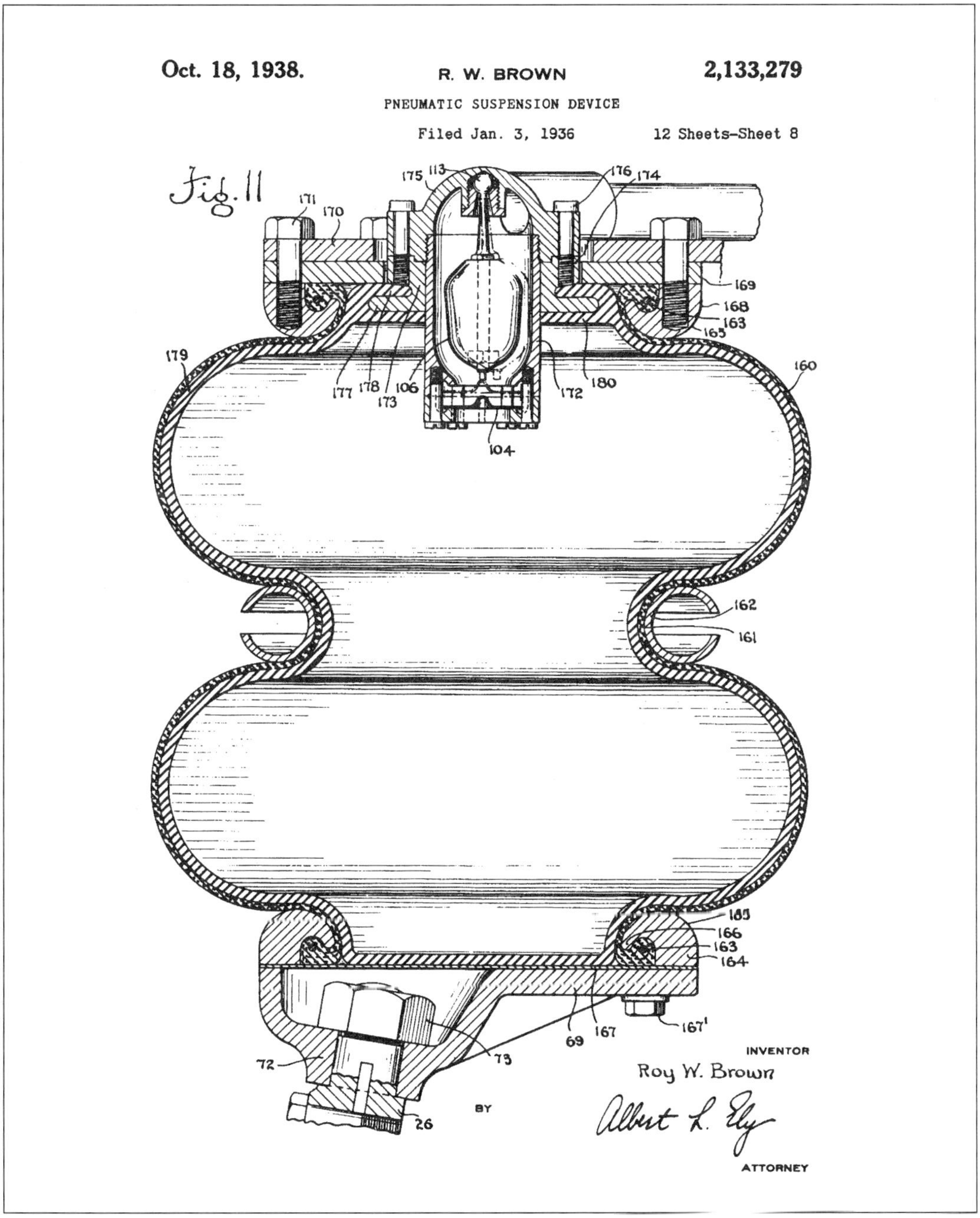

Fig. 2.17 Roy Brown's 1936 patent had 54 claims. The illustration shows his pendulum valve, which throttled air passage to his displacement reservoir when actuated by either lateral or fore-and-aft acceleration forces. The feature was intended to limit roll and to reduce dive upon braking. The valve was also actuated by vertical acceleration forces restricting the flow of air from the reservoir back to the spring—Brown's futile attempt at air damping.

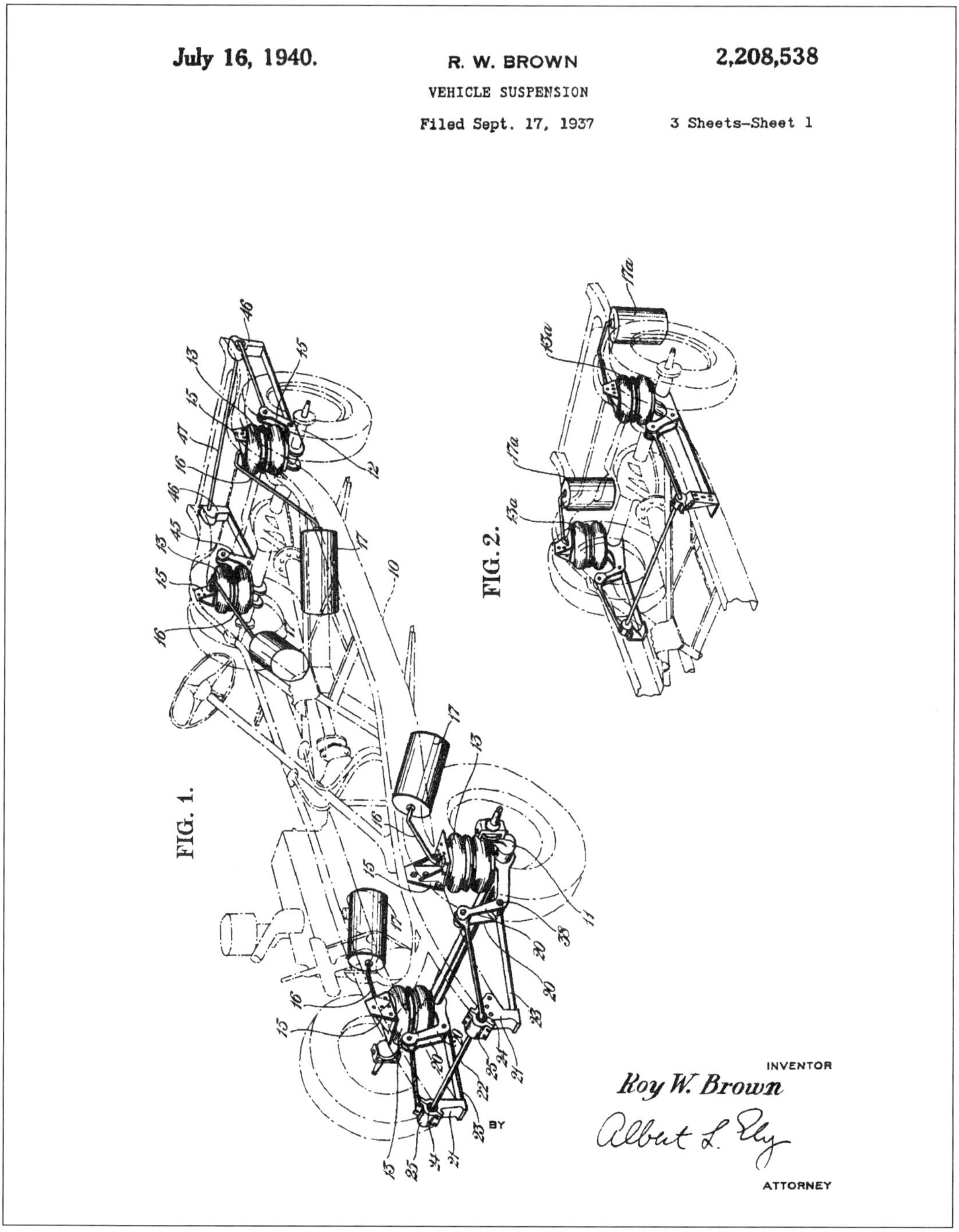

Fig. 2.18 A later Brown patent, showing the layout of the air suspension components he used on his experimental passenger cars. Visible at each wheel is the large-diameter hose that ran from each air spring to its displacement reservoir tank to decrease pressure build-up, reducing the rate of the air spring—which would otherwise be too stiff for a passenger vehicle.

After years of rejection in passenger car engineering circles, Brown began to call on truck, trailer, and bus manufacturers. But Brown was a better engineer than he was a salesman. His manner, at times, could be downright arrogant. Typical of Brown's egocentricity was his habit of tearing pages out of books and technical files in the archives of the Firestone Research Library—a proclivity that eventually got him banned from the company's Central Research Laboratories. He frequently fired employees (sometimes for suggesting that his suspension needed shock absorbers), only to rehire them a few days later.

Seeing beyond Brown's abrasiveness, the chief engineer of the GM Truck & Coach Division, Carl J. Bock, perceived the potential of the technology for his division's trucks and buses—whose changes in payload sometimes exceeded the weight of the empty vehicle. But the United States' entry into World War II abruptly turned General Motors' pursuits to vehicles of a different kind.

In 1947, shortly after veteran rubber engineer Fred L. Haushalter joined Firestone (he had been with B.F. Goodrich before World War II), Bock took up the project once again. Born in 1893, Haushalter had a string of more than eighty patents, dating back to 1922. During World War II he was cited by the army's Chief of Ordnance for his innovative service as a rubber consultant. Radio communication between tanks, for example, would have been impossible without Haushalter's unique method of grounding their rubber-bushed tank tracks, thereby killing the static electricity generated by the otherwise rubber-insulated tracks by sticking needle wires through the rubber bushings.

Carl Bock assigned his research engineer, Harold E. Fox, to work with Haushalter and Brown, who often accompanied Haushalter on early visits to the General Motors Truck & Coach Division's headquarters in Pontiac, Michigan. On more than one occasion, Brown alienated junior engineers, ordering them around with his trademark hubris.

Senior GM engineers William Rice and Donald J. LaBelle soon joined the team to develop what would become a luxurious, radically-new parlor coach with a suspension dedicated to air. Their aim was to solve a problem that had become so serious that it threatened the division's relationship with its major customer, Greyhound Lines.

To reduce the weight of their buses and thereby save fuel, bus designers had gone to all-aluminum bodies in the late 1930s. Greyhound's *Silversides* were beautiful, and they stayed that way because aluminum doesn't rust. But when they were lightly loaded, the handsome coaches lost their glamour. The remaining few passengers got an uncomfortably rough ride on the heavy steel leaf springs which were designed to carry a full load of passengers and luggage; and the fragile aluminum bodies eventually shook to pieces. Even the steel springs broke so often that they might be replaced a half dozen times during the life of the bus (which could easily be fifteen or twenty years).

After five years of design and testing—and redesign and marathon fatigue testing of components in Firestone laboratories (one million full-stroke cycles became the minimum requirement)—assembly of the totally new highway buses began on General Motors' production line in Pontiac in 1952.

The air suspension system is shown schematically in Fig. 2.19.[12] There were a total of eight of Roy Brown's two-convolution air springs in the system. The four on the rear axle (the heavy end of the bus where the diesel engine is located) were about a foot in diameter [30 cm], bulging to almost 13 inches [33 cm] when fully deflected. The ones on the front axle were nominally nine inches in diameter [23 cm], increasing to almost ten inches [25 cm] when fully collapsed. The static design height of both was regulated to a constant 7.5 inches [19 cm] by a system of height control, or "leveling" valves.

Because its diameter changes continuously as it works up and down, a two-convolution air spring (operating somewhat like a round accordion), has an inherent disadvantage when compared to Lewis's or Hoagland's rolling diaphragm configuration of a century

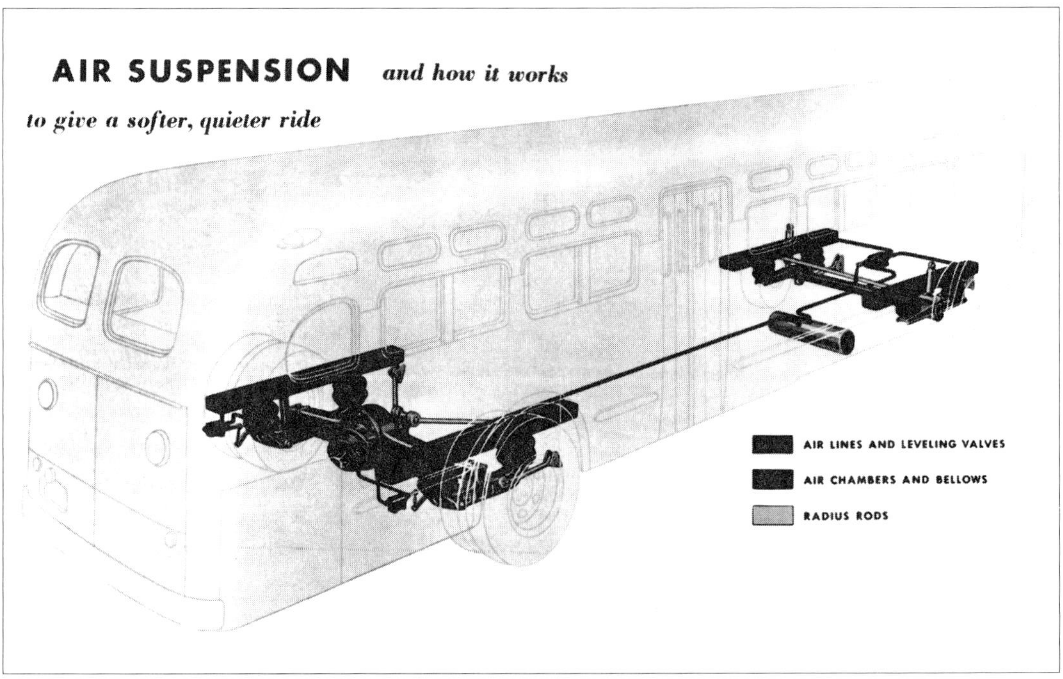

Fig. 2.19 Layout of General Motors Truck & Coach Division's air suspension for both its 1953 parlor coaches and its 1954 transit buses.

earlier—or even a piston in a cylinder—both of which have constant areas throughout the stroke of the spring: As bellows deflect, decreasing the volume and further compressing the air (and increasing the force on the spring) the *effective area* upon which the momentarily higher pressure acts has also increased, further stiffening the spring. The result is a constantly-increasing, exponential load-deflection curve—in contrast to the straight-line function of a coil spring, for example (to be discussed in more detail in the Chapter 4).

As a consequence, if it is closed at both ends, a two-convolution air spring is usually too stiff for a passenger vehicle—which is why each air spring on Brown's converted passenger cars had a large-diameter hose running to a displacement "reservoir" tank to reduce the rate of pressure build-up.

So, on the new General Motors buses, the open tops of each pair of air springs (one pair at each wheel) were clamped by means of bead rings and about a dozen bolts (depending on the diameter of the bellows) to flanges which vented directly into a steel "air beam" reservoir (Fig. 2.20)[8] having a volume (as in the case of Brown's passenger car "reservoirs") somewhat greater than that of the pair of air springs. These beams also served as structural members which supported the bus body. As initially conceived, the air beams were to be lined with a molded rubber bladder, but General Motors production engineers ultimately designed an automated traveling-electric-arc welding system which produced a hermetically-sealed structure. Fig. 2.21 is a side view of the front suspension of the Greyhound bus. The air beam is the horizontal member at the top of the picture.

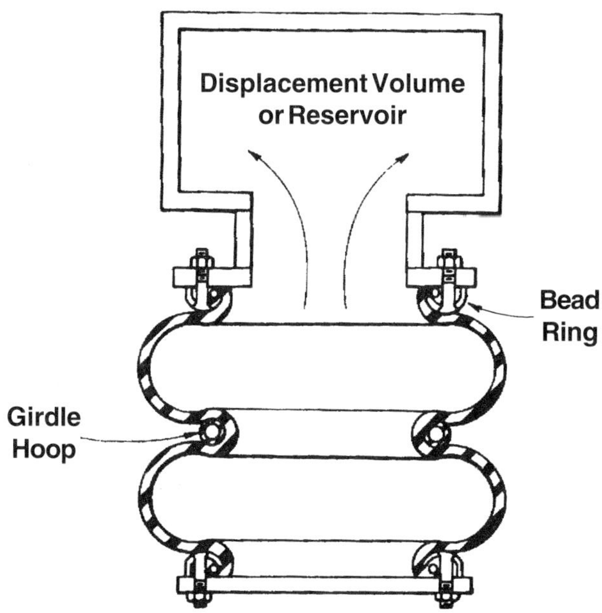

Fig. 2.20 Showing how the two-convolution air spring on GM Truck & Coach Division's air suspension were clamped to a flanged tube which opened into a voluminous "air beam" reservoir to reduce the rate of the air spring.

Fig. 2.21 The first General Motors air suspension. Side view of a 1953 Model 4104 Greyhound bus, showing two-convolution bellows with bead ring and bolt attachment to the flanges of tubular air columns connected to the air beam at the top of the photo. Object in the foreground is the brake drum on the front wheel.

Because the two-convolution GM air springs had relatively little stiffness sideways, steel "radius rods" with rubber bushed ends (Haushalter had been Goodrich's rubber bushing and rubber torsion spring expert) preserved lateral as well as fore-and-aft stability of the axle. And because of the very high elasticity of air (unlike Brown's underdamped passenger cars), the suspension was equipped with (Delco) hydraulic shock absorbers—primarily to control rebound; there was little restriction on the upward jounce stroke.[13]

In cooperation with GM's Delco Division in Dayton, Ohio, Fox, LaBelle, Rice, and Haushalter (Brown had gradually been phased out) designed an elegantly simple system for adjusting the air pressure in the springs as the load in the buses changed. An air compressor kept a set of air supply tanks charged. This compressed air was also used by the air brakes, by the mechanism to open and close the doors on the bus, *and* by the suspension. A line from the secondary "dry" tank led to three "height control" or "leveling" valves. Mounted on the frame of the bus, each valve had a lever—a kind of faucet handle—which was linked to the axle by a vertical rod. There were three of these valves (since three points determine a plane): one at the front axle and two at the rear, one on each side.

As Fox and LaBelle explained in their 1953 SAE technical paper,[14] "As passengers board the coach, the bellows compress, and the body moves down toward the axle. As this happens, the connecting link pushes the valve lever in the proper direction to admit more air to the system. The increased pressure returns the coach body to its original height. Conversely, as passengers leave the coach, the valve lever is pulled in the opposite direction, exhausting air from the bellows [to the atmosphere], and returning the body to its original position with respect to the axle. The height control lever never admits or exhausts air except when the load is changed. It pays no attention to normal movements of the axle.

"Thus, the [spring] rate of the system is always proportional to the load. This results in ... the same soft ride regardless of whether the coach is loaded or empty."[14]

Which is the essence of air suspension.

When Greyhound's new *Airide*®[15] 41-passenger inter-city buses were at last introduced in 1953 (followed several years later by the famous deck-and-a-half Greyhound *SceniCruiser* (also with air suspension), the company could truthfully advertise them as "parlor coaches." Their ride was as smooth as train travel, their reclining foam rubber seats offered the ultimate in comfort, and they even had a lavatory. In 1954, GM brought out a series of all-new, city transit coaches with the same suspension, announcing their "sensational new ride" in a 1953 brochure, as shown in Fig. 2.22. The cover of the brochure (Fig. 2.23) pictured a two-convolution air spring with its bead rings and bolts, with the Delco height control valve seen at the left of the figure.[16]

The structural failure problems that had plagued the *Silversides* virtually disappeared. Greyhound began saving as much as a million dollars a year in bus maintenance alone. In contrast to the former leaf springs, whose life had averaged about 70,000 miles, the new, two-convolution air springs lasted hundreds of thousands of miles. Some actually logged a million miles of service (becoming the subject of yet another Firestone trade journal ad) before they were replaced as a preventive maintenance measure after decades of service.

Early in 1953, in an era before the advent of major air lines offering nonstop flights from New York to Los Angeles, Greyhound ran a double-page advertisement in U.S. magazines (Fig. 2.24), announcing the "biggest news in a quarter century of highway travel: Air-Suspension Ride."[16]

General Motors' introduction of its line of air ride transit coaches in 1954 sparked a revolution in the bus industry. In 1955, Mack Manufacturing Co. (parent company of Mack Truck) brought out a new, 50-passenger city bus for the transit market with a very similar suspension— albeit with the air beam *under* the bellows, mounted *below* the axle,

Fig. 2.22 A 1953 General Motors brochure pictured various models of the new transit buses being offered for 1954—all with air suspension. (John Grafton's Goodyear Collection.)

Fig. 2.23 The cover of another 1953 General Motors brochure pictures the major components of the system: the Delco height control, or "leveling" valve at the left, and a two-convolution Firestone air spring assembly at the right, showing bead rings and bolts for attachment. (John Grafton's Goodyear Collection.)

Fig. 2.24 In an era before major air lines offered nonstop flights from coast to coast, Greyhound ran a double-page advertisement in U.S. magazines in 1953, announcing the "biggest news in a quarter century of highway travel: Air-Suspension Ride." (John Grafton's Goodyear Collection.)

thus reducing floor height—which made it easier for passengers to board the bus. Mack introduced its "97D" inter-city model, having essentially the same suspension, later the same year.

1955 was also the year that Mack bought C. D. Beck & Co., a small bus manufacturer in Sidney, Ohio, and with its purchase, also acquired its own future executive engineer, Donald L. Manning. Two years earlier, in 1953, Beck had introduced its own radically new bus with dual rear axles and a pneumatic suspension engineered by Manning in cooperation with Charles O. Slemmons, then with BF Goodrich Co.

The long, narrow air spring Slemmons had developed for the Beck suspension (one on each side of the rear axle pair) is shown in Fig. 2.25. Approximately 48 inches long by 10 inches wide [25 × 22 cm], the unique configuration had been designed to span tandem axles like these, equalizing the load between them. Essentially an elongated bellows

*Fig. 2.25 Charles O. Slemmons's elongated two-convolution air spring made at the time by BF Goodrich Co., installed on a 1955 C.D. Beck & Co. bus suspension designed by Don Manning. Intended to equalize loads on tandem-axle vehicles, the elongated configuration of the spring spanned the axle pair. To achieve adequate passenger comfort for the bus application, the spring required a huge displacement reservoir, which is shown mounted **below** the spring in the picture. Slemmons later joined General Tire & Rubber Co., which began offering an even longer "Air Pillow" spring for tandem-axle trailers. (Courtesy of Don Manning.)*

configuration, the Slemmons air spring had an effective area characteristic similar to that of Brown's round spring, and therefore required an auxiliary displacement reservoir to depress the natural frequency. Because of the size of the Goodrich spring, in order to achieve a comfortable passenger ride, the Beck suspension required the huge displacement reservoir seen in the figure. It consisted of an 18-inch [46 cm] deep metal box having a width slightly less than the air spring, bridging the 70-inch [178 cm] span between the rear axles. In a typical Manning design decision, in order to lower the floor height of the bus, this reservoir is located *below* the air spring.

After Slemmons and his development team later joined General Tire Co., the group developed a very similar but even longer (66 inch [168 cm]) air spring intended for use on tandem-axle trucks and trailers. In an SAE Technical Paper he delivered in December, 1953, C. O. Slemmons describes his product:

"The spring's construction incorporates two individual cells, placed one over the other, with inter-connecting passageways. The two cells are joined and bonded to the upper and lower metal channels in the curing process. The construction of the cells is the same as that of the sidewall of a tire with the exception that the air envelope, whose counterpart is the innertube, is an integral part of the structure [see Fig. 2.26].[16] The exterior or cover of the spring is made of Neoprene to resist weather cracking, as well as contaminating agents such as petroleum products. The beam is an ordinary weldment, excepting that it must be carefully pressure tested. ... Fittings provide passage of air from the spring to the beam."

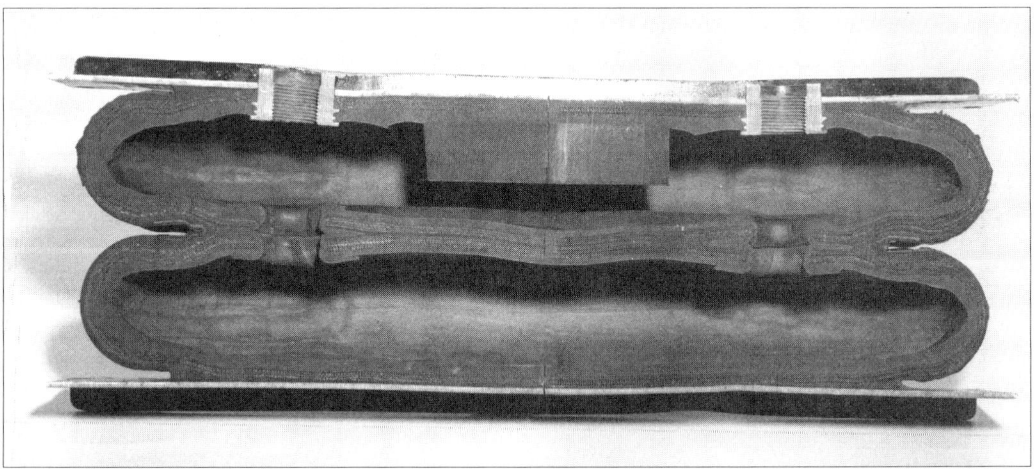

Fig. 2.26 Cross section of the Slemmons (Goodrich) air spring. Slemmons's SAE technical paper explains that it "incorporates two individual cells placed one over the other with inter-connecting passageways. The two cells are joined and bonded to the upper and lower metal channels in the curing process." (Courtesy of Tom Burkley, Goodyear.)

Because this elongated spring had such a large effective area, it operated at a design pressure of only 42 psi [2.9 bar]—less than half that of the design pressure for the Firestone bellows on GM and Mack suspensions.[17]

The Slemmons spring was adopted by several major suspension manufacturers, becoming the Spencer-Stafford "Loadcraft Air Ride," and, after it was adopted by the more successful Hendrickson Company, it became known as the "Hendrickson Air Pillow."

Beck sold 25 of the buses to an operator in Cuba, a sale that resulted in a series of adventures for designer Manning, when the coaches were run over primitive rural Cuban roads, producing nearly eight inches [20 cm] of vertical axle excursion on a suspension that had been designed for a maximum of four inches [10 cm] of articulation. Manning went to Cuba to supervise repairs of the buses, adding further means to limit axle travel.

Manning arrived in Cuba just as the Castro revolution was beginning. The crew on test drives between Havana and Santiago included a shotgun rider as the bus passed through the hills where Castro's revolutionaries were hiding out. And when going into downtown Havana, Manning and his assistant lay on the floor of the taxicab (whose driver got a then-considerable ten dollar tip) in case any of the gunfire they heard was aimed in their direction. The revolution eventually bankrupted the Cuban bus operator, and, salvaging its buses, Beck resold them to Queen City Transit of Cincinnati.[18]

Besides General Motors and Mack, the only other major bus manufacturer at the time was The Flxible Company of Loudonville, Ohio—which, a decade later, would be purchased by Rohr Industries of California. Flxible was the last of the major bus manufacturers to convert its production to air, retaining its Goodrich "Torsilastic" rubber torsion spring (an earlier Haushalter development) until the last.

Marmon-Harrington of Indianapolis converted its production of electric trolley buses to air suspension about 1960. Not long thereafter, Marmon-Harrington was purchased by a South American company.

It was in 1951, while General Motors was gearing up for production, that I first became seriously involved with air springs as a product development engineer in Firestone's Detroit office (and later as an understudy to Fred Haushalter). Earlier, as a junior engineer in Akron, I had known Roy Brown and read his technical papers, and I had even run some tests on his product in connection with a potential railroad application before I was transferred to Detroit. But this hardly prepared me for the psychological environment I would encounter in Pontiac.

It is an understatement to say that veteran engineering and production personnel at both General Motors and Greyhound were very skeptical about the whole concept of air suspension. "Terrified" would be a more accurate description for some of them. And with good reason. To assume the risk of transporting 41 passengers at highway speeds, floating them on a radically new suspension system supported only by "balloons" containing 100 pounds per square inch [6.9 bar] of compressed air, and visualizing what might happen if one of these high-tech infernal devices burst ... It is easy to understand their concern.

To satisfy both Greyhound and General Motors management, the pneumatic system of a prototype Greyhound bus was modified to install a two-inch [51 mm] gate valve on the air beam under the floor behind the driver. A rod with a handle on the end came up through the floor from the left front wheel well. Project engineer John Locklin (with whom I would work for the next fifteen years) drew the assignment of creating the equivalent of a blowout of both left front bellows by pulling the rod.

With Greyhound executives and GM personnel on board and a professional driver at the wheel, the group took the bus down then-rural Opdike Road, a few blocks from the plant. The Greyhound delegation was headed by a short, authoritarian little man who, I believe, was Greyhound's director of purchases—a man named Caesar, who had a reputation for living up to his name. I never learned his first name during the meetings I attended when he was present because I never heard anyone call him anything but *Mister* Caesar.

Beginning cautiously at about 20 miles per hour [32 kph], Locklin gingerly pulled the handle for the first time. Other than a barely-audible "whoosh" of air, and a rapid downward tilt of the left front of the bus (a feature actually built into the right front of transit buses a decade later to assist elderly and handicapped passengers in boarding) the results were anticlimactic. Since the bus had power steering, the driver scarcely felt the effect of the dip.

Gradually increasing the speed, the group tried blowing out the springs at velocities up to 50 mph [80 kph] with Mr. Caesar sometimes calling Locklin's shots, and with the passengers finally gathering sufficient courage to try the experiment at high speed on a curve, all without incident. Both organizations breathed a collective sigh of relief, and celebrated the victory with lunch at an upscale restaurant in nearby Bloomfield Hills.[19]

But the demonstration didn't mute the vigilance of the division's deputy chief inspector and quality control manager (they were separate jobs). As the first production air springs began to arrive, the slightest blemish on the surface was cause for rejection. G. H. Kerber and C. R. Kimler were conservative professionals who had spent their lives in the bus industry, and they had quality standards tattooed on their souls. I remember their first visit to the Firestone Industrial Products manufacturing facility in Noblesville, Indiana. After having dinner with them at an Indianapolis steak house, the evening's entertainment afterward consisted of our standing across the street from the busy Indianapolis bus depot, an old interurban car shed with steel rails still embedded in the floor (in an era when electric interurbans had died, railroads were deliberately giving up passenger traffic, and air travel was too expensive or too scary for many people).

Watching the steady procession of General Motors, Flxible, and Twin Coach buses, Kerber and Kimler entertained us and each other by citing the model number and year of every GM bus as it emerged from the open-sided shelter, reminiscing about the production problems their factory had encountered with each.

Back at the Firestone plant the next day, it seemed reasonable to Kerber and Kimler that their pneumatic springs should never leak any air. So they ordered a one-hundred-percent inspection: every pair of bellows was to be bolted to a pair of end plates, inflated to 100 pounds per square inch [6.9 bar], and immersed in a tank of water to see if any bubbles appeared. They would send their own representative to Noblesville to oversee the operation.

Firestone production engineers quickly tired of manually torquing 18 to 24 bolts on the bead rings of every assembly to be tested, so they designed a fast-acting "octopus" *clamp* which grasped the flats on the bead rings, compressing the rubber bead sealing surfaces against flat steel plates before the assembly was lowered into the large tank of water.

GM inspector Clifford Bauck arrived shortly thereafter to see how things were going. In his role as an authority-figure, Mr. Bauck wore a suit and tie in the sweltering factory heat created by the curing presses on the top floor of the factory. This proved to be a mistake, when, on his first day, intently peering into the tank to see if any "champagne bubbles" appeared around an air spring or at its bead interfaces, the octopus clamp slipped off. With a blast as loud as a howitzer, the pneumatic explosion blew a geyser of water to the factory ceiling, drenching Mr. Bauck's clothes.

Soaked, Bauck walked to the Firestone guest house (the third floor of a Victorian mansion on Connor Street) where he was staying, put on his other suit and, undeterred, returned to the factory. But before the day was over, accompanied by another deafening report heard throughout the plant, the ceiling and Mr. Bauck were awash once again. The event ended Bauck's inspection activities for the day, since he had no more dry clothes.

Immersion testing of every air spring coming off the line would soon be replaced by a more sophisticated nitrous oxide testing apparatus. And General Motors engineers (if not the GM inspectors) would eventually admit that with a copious, continuously-generated air supply that was adequate to operate bus doors and air brakes at every stop, any slight leakage at the bead of an air spring was not really a serious problem. Nevertheless (since that was the only place bubbles were ever seen), the rubber sealing surface of the beads was redesigned and the molds were recut.

The assignment of taking over air spring production from the craftsmen in Roy Brown's Akron laboratory was a traumatic experience for Firestone's Industrial Products Division in Noblesville. Workers were accustomed to molding accelerator pedals, engine mountings, axle bumpers, Maytag wringer rolls, and other familiar mechanical rubber products, including automotive radiator hose, which could withstand up to 12 psi [0.8 bar] of operating pressure. But building a critical, sophisticated new product like air springs (which operated at pressures an order of magnitude higher than the relatively crude, rayon-knit-reinforced radiator hose), on a mass production basis was a new experience for both factory workers and their management. Fortunately, a young mechanical engineer who had been on leave for military service had just returned from Korea to rejoin the company.

Thomas A. Bank had originally been hired by the Firestone Industrial Products Company when he graduated from Purdue in 1949. A veteran of World War II and a reserve officer, Bank had been recalled by the Army to serve in the Korean War, where he flew Cessna L-19s as a field artillery forward observer. He had flown L-4 Piper Cubs (and later L-5s) over France and Germany during World War II—and he had often repaired his own airplanes after "minor" crashes.

When he reported to his former boss, Ed Teegarden, manager of the Noblesville Development Department, manager Ed asked Tom to "see if you could help that group on the third floor. They've got a new product up there they've been having trouble with."

When Tom Bank arrived in the Airidc Dcpartment of the Noblesville factory, he saw a mountain of scrap parts and a host of production problems. The defective rate at the moment was running one hundred percent! Bank would spend the next thirty years solving Airide problems, both in the factory and in customer applications, earning eight patents and the title of Manager of Airide Development, with his own drafting room and technical staff. (The nitrous oxide leak tester mentioned above was typical of his innovations.)

By 1958, Bank's product (with a defective rate now down to two percent) would be moved to a new, dedicated Airide Plant south of the main (mechanical rubber goods) factory, and *Airide* springs would eventually become the only product made by Firestone Industrial Products Company.

As is usually the case with new products, the new bus suspensions were by no means free of problems. One of the earliest—one that became apparent before the buses got out General Motors' factory door—was with Fred Haushalter's rubber bushings on the radius rods, one of which frequently swelled up and popped out of its tapered eye at the end of the sway bars. It was always a bushing in the same location on the left front of the bus.

I did not endear myself to the factory supervisors when, after touring the production line and going down into the pits beneath the partially assembled chassis, I pointed out (and later documented in a letter) a stream of steering gear lubricant trickling down the left front undercarriage of every bus. On most of the buses, this trickle of aromatic hydrocarbon found its way down to the radially-compressed, high-tensile-strength, high-elongation, natural rubber bushings, which had no oil resistance.

A shield, in combination with more care on the part of the factory workers, alleviated the complaint. Usually. But, to Fred Haushalter's dismay, the troublesome bushing (I still remember its part number almost half a century later!)[20] was eventually redesigned as a single double-ended unit which couldn't pop out. The resulting reduction in radial compression may have reduced the fatigue life of the bushings, but it ended the complaints.

In the field, transit buses occasionally had a much bigger problem. Air springs literally tore apart in vertical tension when local bus companies tinkered with the settings of the height control valves to increase the ground clearance of their buses (I sometimes see air springs on highway trailers in this misadjusted condition even today).

Traveling to a transit company's facility in New Jersey in response to the first of these complaints, I could see the cause of the problem as soon as I arrived at the bus turn-around loop—where all of the buses' rubber fenders looked like arched eyebrows in permanent surprise above the tires. But to keep from sounding like a smart aleck upon arrival, I took off my suit jacket and slid under several buses on a creeper, steel tape in hand, before announcing my findings. A GM bulletin limiting the adjusted design height to a maximum of eight inches [20 cm] eliminated the problem.

But the strangest air spring failures we encountered were due to ozone cracking of the rubber inner liner on the *inside* of the bellows. Yet the same Neoprene-natural rubber blend on the covers of the bellows remained intact!

Ozone, O_3, is an unstable form of oxygen which spontaneously breaks down into the familiar O_2 form plus nascent atomic oxygen, O, which is extremely reactive. Natural rubber and most synthetic elastomers crack rapidly in the presence of ozone—producing a distinctive, Y-shaped pattern of long, deep cracks. Atmospheric ozone in most U.S. locations is usually no more than a few (single-digit) parts per hundred million parts of air. After it was discovered that ozone was a significant ingredient in Los Angeles-type smogs (where it can measure up to 90 pphm for a few hours during rare, severe conditions in downtown L.A.), SAE and ASTM testing procedures were modified to increase laboratory ozone test cabinet concentrations from 25 to 50 pphm. But it would take much higher concentrations, we believed, to produce the kind of cracking observed in the bellows interiors.

It was Dr. Glen Mayes of Firestone Central Research Laboratories who ultimately solved the mystery. With portable ozone-measuring equipment he had developed in Akron, Mayes measured concentrations of 100 pphm in the air spring interiors of buses coming off the Pontiac production line. Near the compressor air intake, in the vicinity of the alternator, he measured 39 pphm. But when, on a hunch, he stuck his probe into the exhaust pipe of the bus, the meter went off the scale at 4,000 parts per hundred million!

An analysis of air currents revealed how the ozone was getting into the air system. The air compressor of the bus was downstream from the supercharger of the diesel engine, whose intake was on the side of the block facing the radiator on the rear of the bus. The aerodynamic drag caused by the large, flat end of the vehicle was typically creating a turbulent low-pressure pocket into which the exhaust was being dragged, only to be blown into the engine compartment by the radiator fans. When the air intake was moved to a cleaner location, the interior ozone cracking disappeared.[21]

The potential dangers of diesel exhaust, I later learned, were already well known to women employees of New York Transit, who routinely avoided walking behind buses to avoid having their nylon stockings destroyed by effluent ozone.

One other unanticipated problem—one that arose because of the extraordinary life cycle of buses—was the eventual rusting out of the air beams, deflating the air springs. Blocking off the air beams by interposing a steel plate between the bellows and the air beam, an expedient to which some transit systems resorted, resulted in a natural frequency of over 100 cpm, and a rough, uncomfortable ride for the passengers.

This rusting problem (inevitable in climates where application of road salt was standard operating procedure in winter), was recognized about the same time that General Motors established the new objective of providing *passenger-car ride* for bus passengers on its next model—a goal that led to the invention of a revolutionary new air spring configuration which would offer much lower natural frequencies without any auxiliary displacement reservoir.

The new air spring design would be a Goodyear Tire & Rubber Co. development whose chronology will require more than one additional chapter to document.

Chapter Three

MODEL TRAINS

Shortly after World War II, Robert R. Young, a self-styled financier who owned a controlling interest in more than half a dozen railroads, launched a campaign to reform railroad passenger service in the United States. Encouraging his industry to eliminate rail hub bottlenecks such as Chicago and St. Louis, he ran sensational national ads proclaiming, "A Hog Can Cross America Without Changing Trains—But YOU Can't!" By 1947, draftsmen at the Chesapeake & Ohio Railroad, of which Young had become chairman, began working on what would be called his *"Train X"* design concept, bursting with innovations.[1]

Robert Young's cause got an unlikely assist from a Spanish army engineer, Alejandro Giocoechea. Years earlier, Giocoechea had persuaded Lucus de Oriol, a Spanish philanthropist, that modernized lightweight trains would restore the profits of the Spanish National Railways. A prototype train, the *"Talgo,"* had actually been built before World War II and saw service during the war, collecting data and experience that, it was hoped, would lead to major improvements in a new model to be built after the war. But when that day came, the Spanish railroad industry was incapable of taking on such an assignment.

In 1947, Young asked C&O research director Kenneth A. Browne, to study the *Talgo* train, an assignment Browne turned over to Alan Cripe, a young, visionary engineering graduate from the University of Cincinnati who had become a member of Browne's (and Young's) C&O *Train X* design group. Cripe's input had a profound effect on the C&O's *Train X* design concepts.

Giocoechea turned to American Car & Foundry (ACF) in the U.S. to build the Spanish firm's new *Talgo* train, and ACF brought in Cripe as a consultant to head its *Talgo* design group. Giocoechea's and Cripe's innovative ideas seemed to mesh, resulting in a radically different train than the conservative ACF might have created on its own.

The center of gravity of the *Talgo* coach, for example, was only 3 feet 6 inches [1.1 m] above the rail (in contrast to the American Association of Railroads standard of 5 feet 4 inches [1.6 m]). Cars for the *Talgo* train were less than a quarter the length of American coaches: 20 feet 2 inches [6.2 m] from coupler to coupler; and they would have one axle instead of four—a single set of two wheels per car. There were no vestibules at the ends of the cars. Gaps between the coaches were closed by a zippered, flexible rubber gasket which sealed the sides and roof, giving the interior of the train the appearance of one long tube.

In keeping with Giocoechea's objective of improving performance of the train at high speed on curves, Cripe designed the *Talgo* with air suspension. But the Mechanical Department of the Spanish National Railway refused to go along with this element of the innovative Cripe design, fearing that its personnel would be incapable of servicing such an unfamiliar system. Nevertheless, the seeds of air suspension had been planted in Young's *Train X* design.[2]

Two *Talgo* trains would be built in the U.S. by American Car & Foundry and exported to Spain—where they went into service on July 14, 1950, at a welcoming festival presided over by Presidente Generalissimo Francisco Franco. A third set of six cars was produced for demonstration in the United States.[1]

Having gained from the *Talgo* experience, C&O's design team forged ahead, eventually making static mockups of *Train X* coaches with sealed, perfectly round windows and a strikingly low step height. The mockups were exhibited at the Chicago Railroad Fairs of 1948 and 1949.

To be built of lightweight metals, *Train X* was to have coaches of only a third the weight per passenger of existing trains. Because, as *Railway Age* pointed out at the time, this could result in a weight change of as much as 40 percent, *Train X* would be an ideal candidate for *air suspension*.

The center of gravity of the train would be nearly two feet [0.6 m] lower than American Railroad Association standards—thus improving performance on curves and eliminating the need for step boxes at railroad stations. The suspension linkage would "bank the car into the curves like a pendulum suspension."[3] Self-piloting insert couplings between the cars would include electrical connections which would snap together without manual attention.

Car interiors would be radically restyled to reflect the latest trends in architectural and interior design, and, utilizing new, easily-cleaned materials such as Formica, they would be "passenger proof." The cars would be air conditioned and lighted by tubular fluorescent fixtures (a then-new lighting technology that had emerged just before the war).[3]

By the end of the decade, a prototype *Train X* coach was under construction at the Pullman Standard Company's Griffin, Indiana, Plant. The prototype car was first tested in February, 1951, coupled to a transition car fashioned from a retired caboose, and pulled by a C&O "doodlebug" self-powered coach. Several demonstrations with railroad executives on board were held later that year.

Pullman Standard had developed the pneumatic suspension in cooperation with Roy Brown initially. Later, Firestone engineers Herb Deist and John Diehl recall, they analyzed miles of oscilloscope traces from Pullman Standard test runs in cooperation with a Professor Reissner of Brooklyn Polytech.[4]

As General Motors Truck & Coach Division was gearing up for production of its buses at its Pontiac plant in the early 1950s, a second project was taking shape in its Engineering Department. By 1953, adjacent to the bus production lines, a *train* was being built! And in 1955, the company unveiled its radically new *Aerotrain*, built to Young's *Train X* engineering standards—but with the cars largely based on *bus* technology. It was no accident that the coaches of the train were precisely the length of GM's 40-foot [12.2 m] Greyhound parlor coaches.

The *Aerotrain's* engine, with dramatic "futuristic" styling (Fig. 3.1),[5] was developed by the company's Electro-Motive Division in LaGrange, Illinois, the GM Division that had pioneered diesel-electric technology in the 1930s. In his handsome 1997 book, *New York Central and Trains of the Future*, railroad historian Geoffrey Doughty describes the train:

> "The design for the independent cars was GM's standard 40-passenger aluminum bus body, slightly widened (by 18 inches [0.5 m]) riding on a steel underframe on two single-axle trucks. The cars were modified to use the bus driver's space for an entrance, and the bus engine area for a lavatory and a food service pantry. As in the bus, luggage was carried in 150-cubic-foot compartments located under the floor section, accessible from outside the car. [These are visible in the side view of the train, in Fig. 3.2; and the use of modular bus components is apparent in the *Aerotrain* engineering drawing presented in Fig. 3.3.][5]
>
> "… Each train set featured a futuristic observation car which looked like a cross between the trunk of a 1956 Cadillac and something out of the early 1950s science fiction classic, *The Day the Earth Stood Still*. They sported white and red tail lights with the letters 'GM' between them, as delivered."[7]

(The Pennsylvania Railroad, however, covered up the GM logo with its red keystone herald.)

Fig. 3.1 The General Motors ***Aerotrain****. The diesel-electric engine was produced by the Electro-Motive Division; air-suspended coaches were made by GM's Truck & Coach Division. (General Motors Electro-Motive Division.)*

Fig. 3.2 The coaches for the ***Aerotrain*** *bear a strong resemblance to General Motors' parlor coaches, whose bus components were used extensively. Baggage for the rail passengers was stored in compartments beneath the floor, accessible only from outside the rail car.*

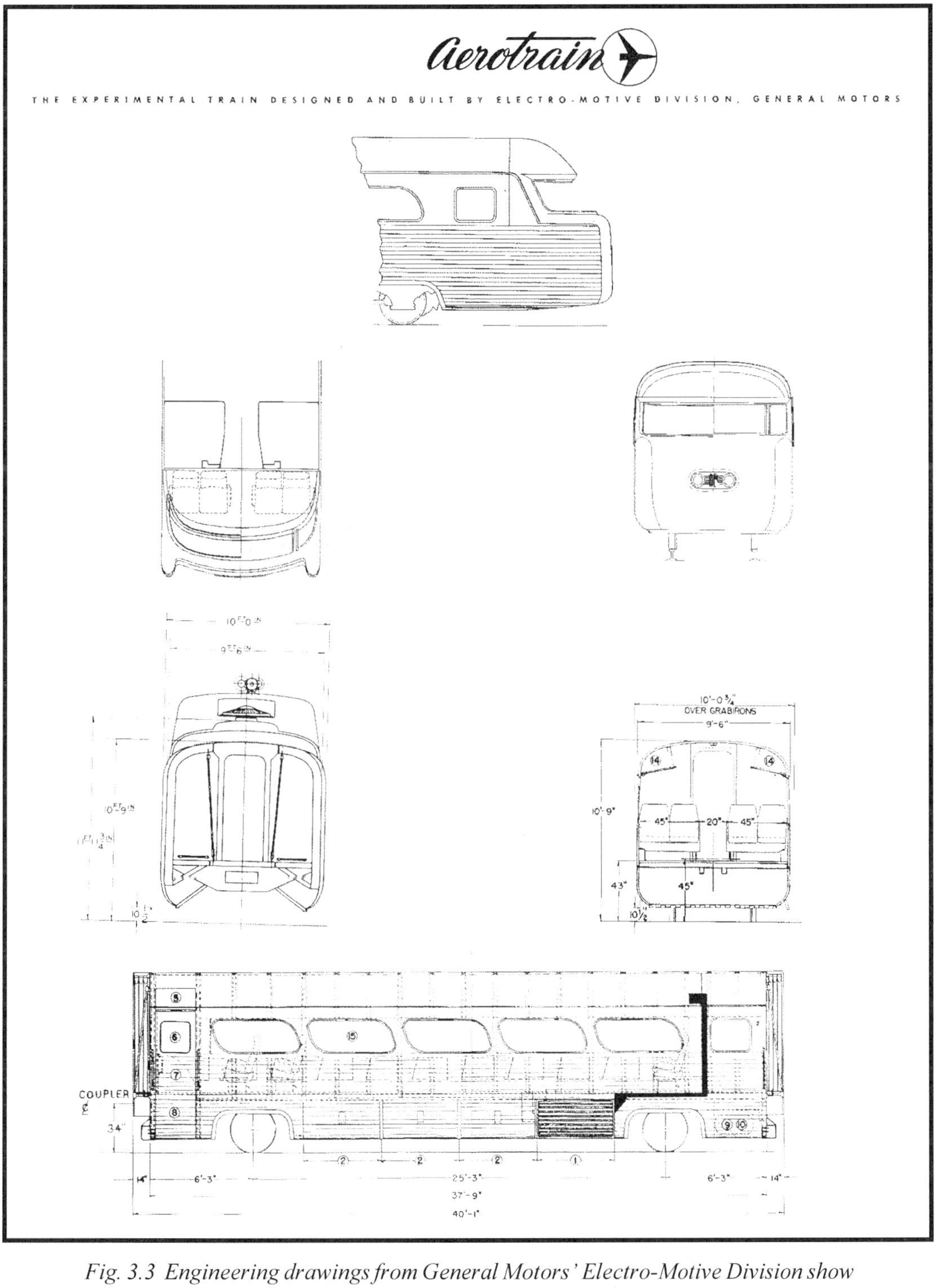

Fig. 3.3 Engineering drawings from General Motors' Electro-Motive Division show how the 40-foot bus bodies were modified to become rail cars.

The air suspension of the train, too, was similar to that of the GM buses, but with larger-diameter, *three*-convolution bellows—four per axle. (A three-convolution bellows, similar to the ones used on the Aerotrain, is shown in Fig. 3.4.) In place of the air beams on the bus chassis, the bellows displaced air into a metal box which served as a displacement reservoir (Fig. 3.5).[6]

Fig. 3.4 A three-convolution bellows similar to the ones used on ***Aerotrain*** *coaches (four per axle) on a laboratory durability machine. (Tom Burkley, Goodyear.)*

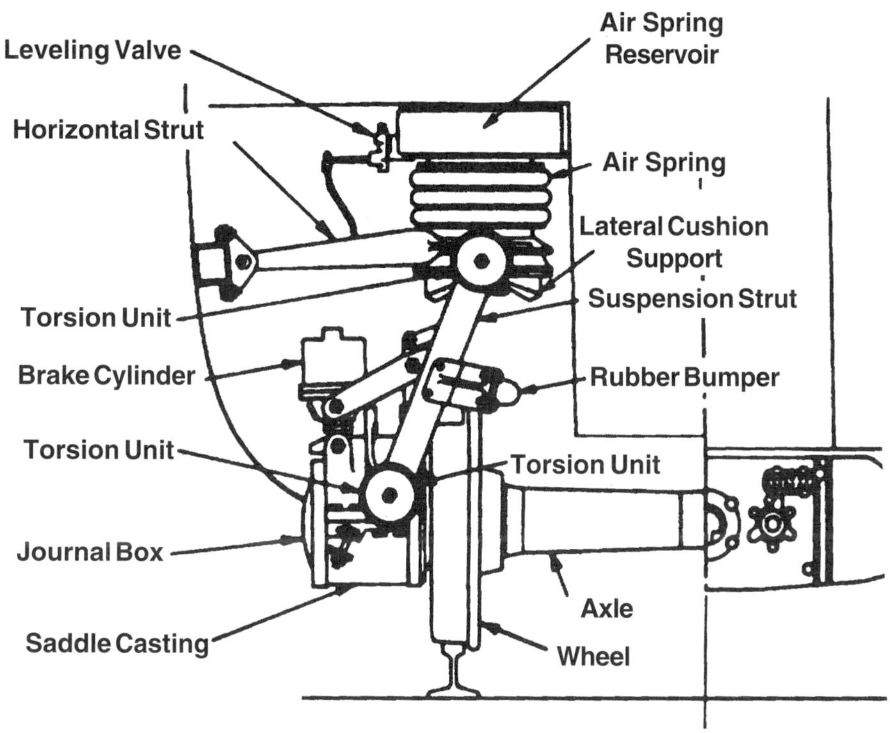

Fig. 3.5 The geometry of the radius rod linkage used on buses was altered to make the train bank into a curve. As the centrifugal force caused the body to go in the direction corresponding to the left in this illustration, it pulled the horizontal strut to the outside of the curve, causing the vertically-slanting radius rod to move through an arc—upward and to the left, raising this side of the car and leaning the coach toward the inside of the curve. (Railway Age.)

The geometry of the radius rod linkage was altered from that on the buses in order to make the train bank into a curve. This feature was *not* produced, as on some later trains, by increasing the pressure in the outboard bellows. Rather, with reference to the above figure, as the centrifugal force causes the body to go left, pulling the horizontal strut to the outside of the curve, it causes the nearly vertical radius rod to move through an arc upward and to the left, raising this side of the car and leaning the coach toward the inside of the curve.[3]

Perhaps inspired by the "planned obsolescence" in vogue in the automotive industry at the time, GM's train was designed so that, instead of car interiors receiving a customary overhaul and renovation after seven years, the entire coach bodies could be lifted from their chassis and attendant running gear and scrapped—to be replaced by new bodies from GM's Pontiac factory (which could then incorporate "appropriate improvements and styling changes").

General Motors produced two trains. *Aerotrain* 1000 was delivered to the Pennsylvania Railroad in February, 1956, and, after a month of revenue service, made a circuit of six other railroads. No. 1001 went first to the Santa Fe for a week with a rail-fan trip from Chicago to Los Angeles and an exhibit there; then on to the Union Pacific, Southern Pacific, and four other roads.

Although they were designed as ten-car entities, the trains usually ran with eight coaches. One car of each was used by GM engineers for further testing—especially *ride* testing, a property that turned out to need a great deal of tuning, despite the advent of air suspension. The other car for each train was kept by the railroads as a spare.

By the time Pullman Standard brought out its *Xplorer* in 1956, the round windows on its *Train X* prototype had become rectangular, and the four wheels per coach on the competitive *Aerotrain* had been reduced to two. In a radical departure from conventional coupling technology, the cars hooked together like the elephant-train trick seen in circuses. But the single pair of wheels was at the *front* of the coach with the hitch at the rear, the weight of the rear end of each forward coach thus resting upon the axle of the coach behind. The *Xplorer* went into service on the New York Central, and Pullman made a second "non-identical twin," the *Daniel Webster*, which was also equipped with air suspension.

Firestone took advantage of the opportunity to publicize its participation in the *Train X* experiments in a May, 1956, advertisement in *Railway Age* (Fig. 3.6), using the opportunity to extol the virtues of air suspension at the same time. Typically, the fine print of the ad reflected the advertising copywriter's fascination with the medium instead of the functionality of the suspension. It began (sounding like the text of a nineteenth-century patent) "Firestone has imprisoned the softest of all things, air. ... Train X provides a buoyantly, comfortable shock-free ride. Passengers really *ride on air.*"

Unfortunately, as implied earlier, the ride on the X trains was not that spectacular. Pullman Standard's air suspension system had the advantage of being a second generation—having been first worked out in the design of the company's *Train X* prototype. As a consequence, the ride of the *Xplorer* got slightly better marks than the *Aerotrain* when passengers filled out rating forms. But none of the new trains lived up to expectations. They "jiggled more like a bus than a train."[1] Noise levels were much higher than train passengers were used to. The cars had a better ride at higher speeds than when going slower (possibly because the rapid "clickety-clack" wheel-rail input associated with high speeds was easier to isolate than the low, rumbling frequencies produced at lower speeds). Furthermore, according to a General Motors test document, at 100 miles per hour [161 kph], "the laterals were terrible."[1]

Fig. 3.6 Advertisement published by Firestone in the May 7, 1956, issue of ***Railway Age****.*

Probably because of large-excursion oscillations permitted by the suspension, the *Aerotrain* derailed the first time it entered Grand Central Station! Even worse, precipitated by an electrical problem, the *Daniel Webster* caught fire on its initial press run.

Many trips back to the manufacturers improved the ride, the jiggle, and noise levels of both the General Motors and Pullman Standard X trains. But, like the passenger car adventure to follow, the X trains were to be a major disappointment for advocates of air suspension. One *Aerotrain* was scrapped in the mid 1960s; the other became an exhibit in the National Transportation Museum in Green Bay, Wisconsin, in 1967. The *Xplorer* was sold to James F. Jones, president of the Pickens Railroad (as was New Haven's *Daniel Webster*) in May, 1960, and was subsequently employed by Jones in tourist excursion service. In 1963, it was used in tests by United Aircraft during development of the *Turbotrain*.[7]

There were other forays into experimental trains, including the Budd Company's *Pioneer III* (a refinement of its 85-foot [23 m] stainless steel coaches, but with a conventional railroad suspension). And there was United Aircraft's *Turbotrain*, which was designed by Alan Cripe. Like all of Cripe's creations, it had air suspension. But none of the new trains blazed a trail of success.

Entirely aside from the problems detailed above, there were a number of factors that presaged the doom of railroad passenger service at the time. Having been spoiled by gas and tire rationing and gratuitous profits from troop trains during World War II, the railroads were unprepared for the postwar world they would encounter after the war ended. With the freedom to drive their cars once again and, for many, enough income to buy a car for the first time (and discovering the glorious independence a personal car confers), there was a natural exodus not only from train travel, but from city transit systems as well.

Travelers who analyzed the relative cost, quickly discovered that it was actually *cheaper* to drive their own car to a distant city than to take the train; and, of course, they would then have local transportation when they got there.

Instead of creative marketing programs to lure their passengers back, railroads attempted to make their passengers fit the needs of the railroad instead of the other way around. In the early 1950s, when I occasionally had lunch in the Michigan Central Building near the Firestone office in Detroit, I can remember overhearing conservative railroad executives at nearby tables complaining about their lower-profit passenger trains "getting in the way of freight trains, where we make our money."

There were other financial pressures. Left over from curbs on America's railroads instituted in 1929 following the era of the robber barons, the Interstate Commerce Commission limited the return on railroad investments to three percent. Such was not the case with airlines, which were achieving more than twice the return of railroads at the time (sometimes with the aid of government subsidies). There followed a period of bankruptcies and wholesale buy-outs—which is one of the reasons that Young personally controlled more than half a dozen railroads. The world had changed, but the railroads hadn't.

From an engineering perspective, I believe that there are several reasons for the disappointing ride performance and noise levels of the X trains. First, their coaches weighed only about a third as much per passenger as did the heavy, solidly built traditional rail cars, which were not only more soundproof but were also free of tinny rattles. Second, I believe that there is a natural tendency on the part of passengers traveling on buses over uneven roads (this was before the age of interstates) to expect and excuse a bit of muted "jiggle" to which they are not accustomed on the relatively smooth rails traversed by trains. But, most important, I think that the relatively high natural frequencies of the air springs on the X trains—ride frequencies that produced a comfortable ride on buses—which had the luxury of *pneumatic tires*—could not be expected to be adequate for vehicles whose steel wheels were constantly impacting irregularities in steel rails. On buses, the tires provide a second set of pneumatic springs in series with the suspension springs of the vehicle, and *ahead of the unsprung mass.*

When air springs having a substantially lower natural frequency than multiple-convolution bellows were introduced a decade later, they overcame these problems sufficiently to capture the light-rail mass transit market, and they led to widespread standardization of air suspension on high-speed rail in Japan, in Europe, and eventually in the United States.

But to keep this chronology in historical order, we must first recount the adventure of the automobile industry's first attempt to apply air springs to production passenger cars. However, before we do that, in order to understand some of the problems that arose during that traumatic period, it is necessary to explore some basic technical issues, including how suspension performance affects ride comfort, and, more important, the *physics* of pneumatic springs—how fickle Gas Law properties and the nuances of thermodynamics govern, and sometimes confuse, the performance of air springs.

Chapter Four

COMPLYING WITH THE GAS LAWS

This is a difficult chapter to write because, on the one hand, the principles of engineering mechanics outlined in the next few pages will be common knowledge to engineers raised on a diet of statics, dynamics, resistance of materials, and thermodynamics. On the other hand, I don't want to lose historians, MBAs, or other readers who haven't had this (sometimes doubtful) intellectual pleasure. But understanding the basic physics of air suspension is necessary to an appreciation of what makes air springs unique. So I ask the indulgence of both groups as they wade through (or skip at their option) the material that follows.

After more than a dozen years of experiments in his laboratory at Oxford, incidentally inventing the mechanical vacuum pump in order to carry out his studies, the prescient Anglo-Irish physicist, Robert Boyle, published a paper in 1660 having an interesting title from the perspective of this book: "New Experiments—Touching the Spring of Air."

Two years later, in 1662, he announced what has come to be known as Boyle's Law: *The volume of a confined body of gas varies inversely as its absolute pressure, provided the temperature remains unchanged.* (Absolute pressure is the *total* pressure of the gas, i.e., gage pressure plus atmospheric pressure, or 14.7 psi [approximately 100 N/cm^2 or one bar].)[1] A century later, French scientist Jacques A. C. Charles concluded that if the volume of a gas remains unchanged, its *absolute pressure increases directly with its absolute temperature.* (Absolute temperature is measured from absolute zero, which is –459°F, or –273°C.)

A simple formula for the General Gas Law relating pressure, temperature, and volume, one that appears in physics books, may be written:

$$\frac{P_1V_1}{T_1} = \frac{P_2V_2}{T_2}$$

In which P_1 = Initial pressure (absolute)

P_2 = Final pressure (absolute)

V_1 = Initial volume

V_2 = Final volume

T_1 = Initial temperature (absolute)

T_2 = Final temperature (absolute)

For me, visualizing the dynamics of gas behavior on a submicroscopic gas-particle level makes it easier to understand the intimate interrelationship between the volume, pressure, and temperature of a gas. In air, a mixture of gases, single molecular particles are in constant motion, bouncing around furiously in the atmosphere—or even more furiously inside the closed container of an air spring because the pressure is higher.

Pressure is a measure of the average *impact force per unit area* that is produced by these particles hitting against the walls of the container (or striking the diaphragm of a pressure gauge). When the gas is compressed, these particles become more numerous per unit volume as they are squeezed more tightly together, and, because they hit the walls of the container more frequently, this registers as higher pressure.

Temperature is a measure of the *energy* of these particles. Their (average) velocity changes with temperature. The hotter it gets, the faster they move and bounce. And since this causes them to hit the walls more frequently, we measure this as increased *pressure.* (At extremely low temperatures, the particles quit bouncing altogether, settling on the bottom or sides of the container, condensing as liquids or crystallizing into solids.)

Thus, when the volume of an air spring is reduced during the jounce stroke of a suspension, confining the particles more tightly together, this not only increases the pressure (the number of collisions against the container walls), but also the temperature (the average velocity of the particles), because the energy imparted in the squeeze has been transformed into increased kinetic energy of the particles—which is measured as *heat.*

Because (consequently) all gases heat up when compressed and cool when expanded, in order for an air spring to behave in accordance with Boyle's Law, it would have to be stroked so slowly that all of the heat generated during compression (or absorbed during expansion) was dissipated, so that its temperature remained unchanged—what is called an *isothermal* (constant temperature) process. In isothermal operation, for an incremental amount of travel (into either the upward compression or "jounce" stroke, or the downward "rebound" stroke) the formula for the resulting new pressure is very simple:

$$P_2 = P_1 \frac{V_1}{V_2}$$

In which P_1 = Initial pressure (absolute)

P_2 = Final pressure (absolute)

V_1 = Initial volume

V_2 = Final volume after compression or expansion

Volumes in this expression are *total* volumes. If the air spring is connected to a displacement chamber, such as the "reservoir" on Roy Brown's passenger car suspension, or it vents into an "air beam" as on General Motors' early buses, the fixed volume of this chamber must be included, added to both the initial volume of the air spring and the final volume after deflection.

In the real world, as a vehicle is driven over the road with continuous working of the suspension and attendant flexing of the air springs, isothermal performance is impossible. During the compression stroke, very little of the heat of compression can escape—and the interior cannot instantaneously rewarm after expanding during rebound. Thus, during compression, the temperature is further increased in accordance with Charles' Law; and, as the jounce stroke of the air spring reverses and goes into rebound, the sudden decrease in pressure causes the internal temperature in the air spring to instantaneously drop. If a pneumatic spring is stroked fast enough that all of the heat of the operation is conserved, i.e., if an *adiabatic* (or constant heat) process occurs, the above expression, for air, becomes:

$$P_2 = P_1\left(\frac{V_1}{V_2}\right)^{1.4}$$

The value of the above exponent (the gas constant *gamma*) varies with the gas, and is a function of specific heat of the gas.[2] The gamma value for helium is 1.66; for carbon dioxide, 1.30; for sulfur dioxide, 1.24; and for butane, 1.09.

Recognizing the rate at which suspension movements take place, the *ride frequencies* typical of air springs (see below), Roy Brown correctly decided that the action was not quite adiabatic, and arbitrarily established the convention of using 1.38 as his gamma value for air spring calculations. For many years thereafter, before the availability of pocket calculators, Firestone's Airide Design Manual included a thick appendix of tables to help engineers raise numbers to the 1.38 power if they didn't happen to have a loglog decitrig slide rule. That this "constant" in practice is arbitrary is evidenced by the fact that, within General Motors, GM Technical Center engineers used 1.4 for the gamma exponent while Buick used a value of 1.3 when calculating values for the same air spring. And both Firestone and Goodyear used 1.38.

The above equation became the standard formula for calculating the load-deflection curve (point by point, with hand plotting in years past) for a proposed air spring, before a mold was ever made—provided the *effective area* of the spring could be estimated incrementally during spring travel. Effective area is the load-carrying area of a pneumatic spring at a given point in the stroke. Phrased the other way, it is the load on the spring divided by the air pressure in the spring. For a piston in a cylinder the effective area of the spring is constant, but the effective area of a flexible air spring changes significantly during its stroke. Indeed, tailoring this pattern of change is an important part of air spring design—a nuance recognized as early as 1910 by inventor Benjamin Bell (Chapter 2).

To determine the force, or load, at any given point in the stroke of an air spring, atmospheric pressure of 14.7 psi (one bar) must be backed out of the absolute pressure values to yield *gauge pressure*. The force, or load on the spring at any given point on the load-deflection curve, thus, is the *gauge pressure multiplied by the effective area.*

In a flexible reinforced-rubber air spring, the effective area is a function of the instantaneous *effective diameter* of the spring—which is the distance between centers of the radii of the curvature of the bulging lobes of the air spring (which, in cross section, nearly always approximates a segment of a circle). Fig. 4.1 shows a typical two-convolution air spring (with a crimped-on bead plate sealing the top and the bottom). Fig. 4.2[3] illustrates how the radius centers of the inflated lobes of this air spring move apart as the spring is compressed, increasing the effective area, and how they move together toward the center as the spring extends.

Fig. 4.1 A commercial two-convolution air spring with crimped bead plates sealing the ends and a cured-in girdle hoop (both developments by Tom Bank). The bead plate has blind nuts and an NPT air fitting.

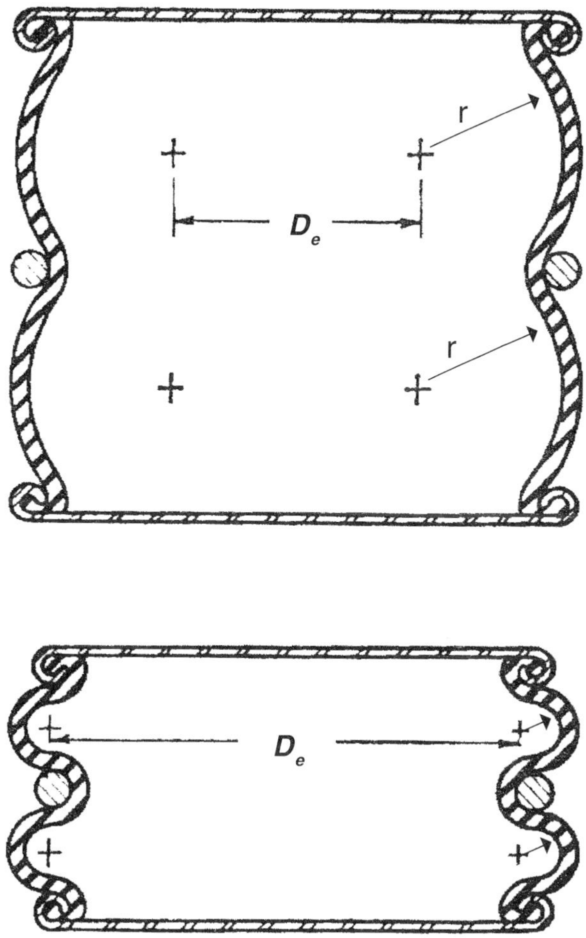

Fig. 4.2 Illustrating how the effective area of an air spring varies as the bulge of the convolution changes shape: increasing the radius of the sidewall as the bellows lengthens from design position (lower sketch) into rebound (upper picture), causing these radius centers to move toward one another, thus decreasing the effective diameter. Collapsed into full jounce, as these radius centers move farther apart, the effective diameter becomes greater than at design position.

At the risk of overdoing this explanation to clarify the concept: after a basketball is pumped up, these radius-center points are coincident in the center of the sphere as long as the ball remains undeflected. If the ball is stepped on, however, distorting it into a spheroid, the radius centers will separate, moving apart just enough that the effective area thus formed (an imaginary disc whose diameter is bounded by the separated radius centers of the new pumpkin shape) multiplied by the new pressure will exactly balance the weight placed upon the ball.

Shown in Fig. 4.3 is a typical effective area vs. deflection curve for a two-convolution air spring (read right y-axis for area) together with its resulting load-deflection curve (left y-axis) without any auxiliary volume. Fig. 4.4 shows a pair of load-deflection curves for a (smaller size) two-convolution air spring, illustrating what happens to the load-deflection curve as its pressure is adjusted by the leveling valves to accommodate changes in vehicle load.

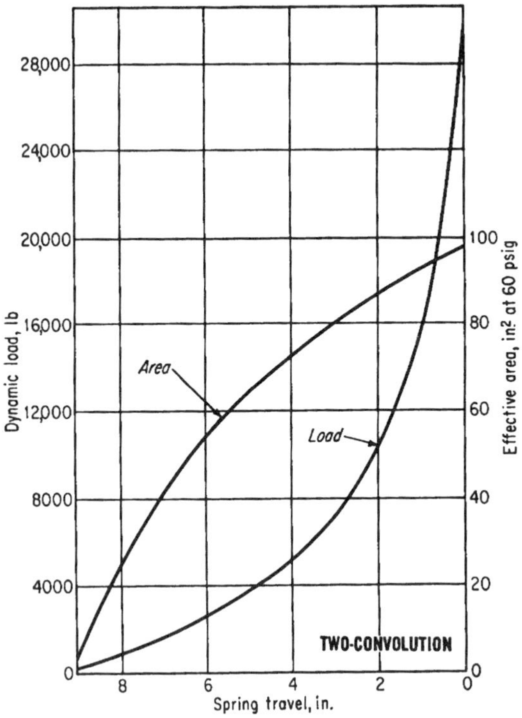

Fig. 4.3 Curve shows how the effective area of a two-convolution air spring changes with stroke, resulting in the characteristic exponential shape of its load-deflection curve.

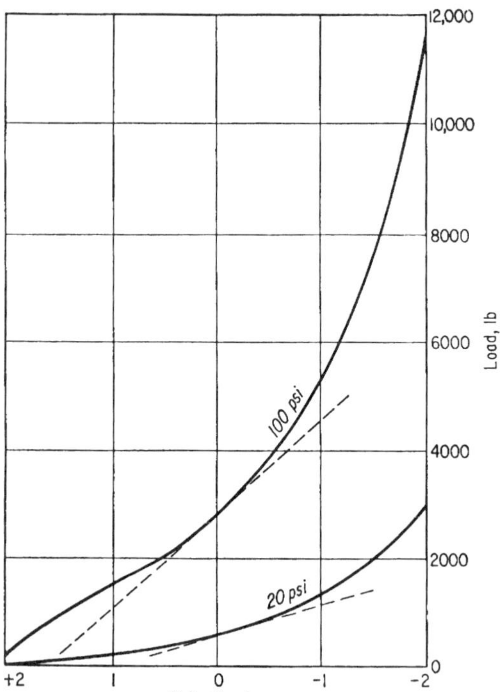

Fig. 4.4 As the pressure in a two-convolution air spring is decreased, its load-deflection curve not only moves lower on the chart, but the slope of the curve, a measure of its spring rate (see tangential dotted lines at the "zero" design position) has decreased in proportion to the load—thus maintaining a constant k/m ratio and preserving a constant natural frequency regardless of the load.

Shown at the (zero-deflection) design position for each pressure is a dotted line tangent to the curve. The *slope* of this line (in lb/in. or N/cm) is the *spring rate* at that spring height. When load is added to a vehicle equipped with air suspension and the height control valve admits more air to the springs to restore the vehicle's design height, the increased air pressure required increases the slope of the tangential dotted line (stiffens the spring) in direct proportion to the load added. Preserving the ratio of the spring rate to the vehicle's mass results in an essentially constant *natural frequency*—one of the primary advantages of air suspension.

When any mass, suspended from or resting upon a spring, is excited in the vertical mode, it will bounce up and down at the resonant or "natural" frequency of the individual spring-mass system. The frequency rate of this undamped vertical harmonic motion, the number of times the mass will bounce up and down per unit time (usually expressed in *cycles per minute* for vehicle suspensions), can be calculated accurately with a relatively simple formula:

$$F = \frac{1}{2\pi}\sqrt{\frac{k}{m}}$$

In which F = Natural frequency

k = Spring rate

m = Mass resting on the spring

For springs having a flat, linear load-deflection characteristic, such as coil springs, this formula may be abbreviated (in English units) to:

$$F = \frac{188}{\sqrt{d}}$$

In which **d** is the static deflection of the spring (expressed in inches for the constant shown, with **F** in cycles per minute).

This shorthand formula can be used for springs with more complex, nonlinear load-deflection curves, including air springs, simply by dividing the load at a given point on the curve (usually the static load), by the spring rate, or slope of the curve at that tangent point, yielding a functional *effective static deflection*—which is unrelated to the physical dimensions of the spring, but which can be plugged in for **d** in the denominator of the expression.

Natural frequency is a fundamental value which must be known to predict the efficiency of a spring system's ability to *isolate* disturbing vibrations. The *transmissibility* of a disturbing (or "forced") frequency (F_f) to be isolated is inversely proportional to the natural frequency (F_n) of the system, and may be calculated according to the relationship:

$$T(\%) = 100 \div \left(\frac{F_f}{F_n}\right)^2 - 1$$

It is apparent that the natural frequency of the isolating system must always be substantially lower than the disturbing frequency to be isolated. Indeed, if the two are nearly equal, the disturbing frequency will be *amplified*, as becomes clear when natural frequency is plotted against the disturbing (or "forced") frequency (see Figs. 4.5 and 4.6).

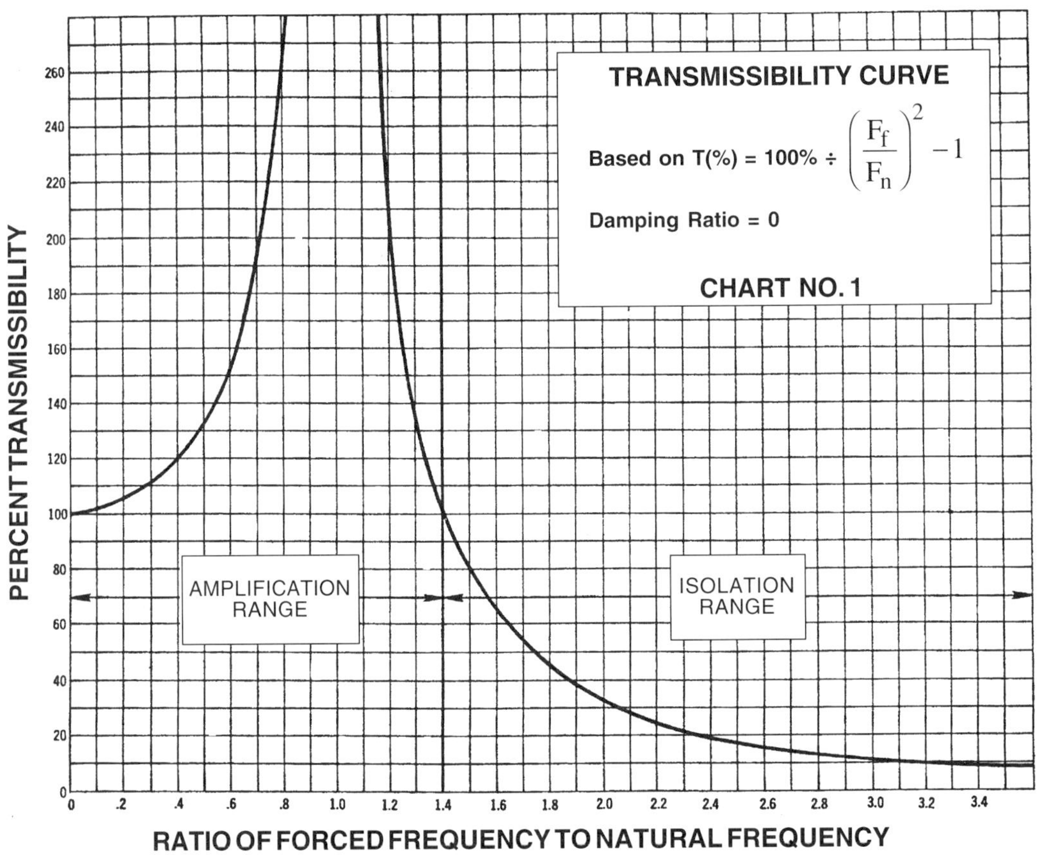

Fig. 4.5 When a mass is mounted on a spring to achieve vibration isolation, the transmissibility of the vibration to be isolated is inversely proportional to the ratio of this disturbing or "forced" frequency to the natural frequency of the system, i.e., the lower the natural frequency, the greater the isolation. Transmissibility increases as this ratio decreases, causing amplification of the vibration as the ratio approaches unity.

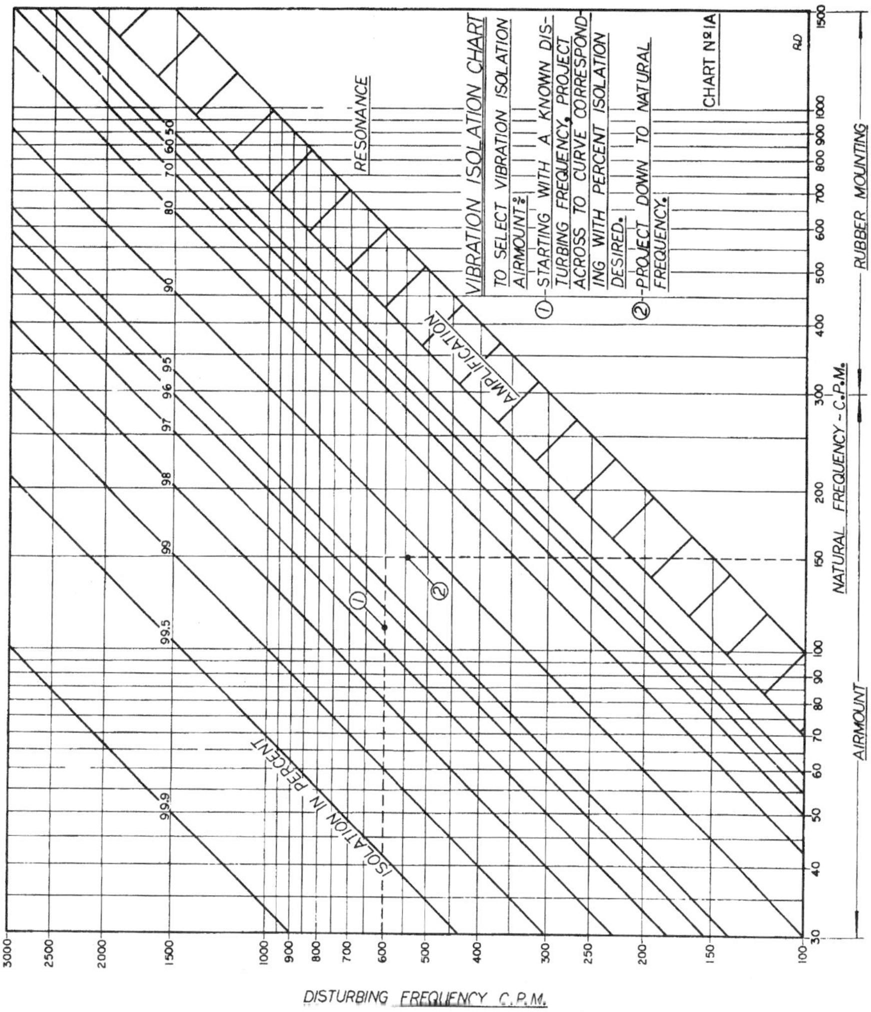

Fig. 4.6 A chart for graphically estimating the degree of vibration isolation to be achieved when the values of the natural frequency and disturbing frequency are known; showing how air springs, having the potential of a significantly lower natural frequency, usually can achieve more efficient isolation than rubber mountings.

To demonstrate how these charts can be used in a practical application: Let us assume an electric motor with a slightly unbalanced shaft that turns at 1800 rpm. With reference to Fig. 4.6, it will be seen that putting rubber feet (or coil springs) that deflect 1/16 in. [1.6 mm] under the feet of the motor will create a mounting system that produces a natural frequency for the sprung mass of 800 cpm. The rubber feet will result in 80 percent isolation (the reciprocal of transmissibility). Softer mounts increasing the deflection to 3/16 in. [4.3 mm] will drop the natural frequency to 450 cpm, and will increase the isolation to 95 percent. But mounting the motor on much softer air springs having an *effective* static deflection of 3.5 in. [90 mm], producing a natural frequency of 100 cpm, will achieve 99.7 percent isolation.

As stated above, the effective static deflection of an air spring has nothing to do with the dimensions of the product itself. An air spring can have an effective static deflection greater than its own physical height, making air springs convenient as well as effective in a wide variety of practical isolation applications including air conditioning equipment in high-rise office buildings, sound booths, Navy undersea research vessels, the Space Shuttle, tanks in which germanium crystals are grown, and many other vibration-sensitive applications.

Getting back to vehicle suspensions, the subject of this book: Oversimplified, it follows from the above that the lower the natural frequency of a suspension, the more comfortable the vehicle ride is likely to be, that is, the more isolation from a bumpy road the suspension can be expected to provide. But the natural frequency must not be so low that it contributes to excessive roll or pitch and consequent motion sickness. And the system must be *damped* with shock absorbers.

Although natural frequency in the vertical mode is widely used as a criterion for suspensions, it has been established by quantified subjective testing that the human comfort index is actually *least* sensitive to vertical input—perhaps because walking desensitizes us to it. We are significantly more sensitive to fore-and-aft shocks, and *most* sensitive to lateral accelerations (which are the most likely to create motion sickness). But because, in the real world on real roads, much more *energy* is transmitted to the passenger vertically than in either of the other modes, the vertical mode is the one primarily addressed in the design of vehicle suspensions.

One problem with conventional steel springs is that the natural frequency does not remain constant. A ten-ton truck with a tare weight of five tons can be designed to have a comfortable ride when fully loaded, but it will "ride like a truck" when empty because it will have lost half its mass *and* its spring deflection, shooting up the natural frequency accordingly. Furthermore, even on a vehicle whose weight does not change significantly, a very soft suspension that yields a comfortable ride on relatively smooth roads will continually slam into the suspension bumpers at both ends of the spring travel when the vehicle is driven on rough roads or on roads with dips in the pavement—depending, however, on the shape of the load-deflection curve and the shock absorbers in the system.

Air suspension proved to be an answer to this long-recognized problem, as pointed out in Don Labelle's SAE technical paper (see Chapter 2). A pneumatic system with automatic height control, or "leveling valves," offers an essentially *constant natural frequency* regardless of thc load. When weight is added to the vehicle and additional air is pumped into the spring to restore the design height, the pressure is increased. Because the spring rate of an air spring, ***k***, is a function of the absolute pressure of its contained air, the ***k*** in the above natural frequency equation goes up in proportion to the mass, ***m,*** added, resulting in a *constant k/m ratio*, i.e., a constant natural frequency—a constant *ride*. This unique feature permits an air suspension designer to select as low a natural frequency as is consistent with the performance demands and terrain to be negotiated by the vehicle. It would, for example, be impractical for an off-road mining truck to have a natural frequency as low as that of a bus or a passenger car.

Without an auxiliary-volume displacement reservoir, the natural frequency of all two-convolution air springs (Fig. 4.1), when inflated to their design position, regardless of load, will be in the range of 100–115 cycles per minute (cpm). (Three-convolutions drop this value into the 80–90 cpm range). Single convolution air springs have a natural frequency of 150 to 200 cpm or higher. (Newer rolling lobe and rolling diaphragm designs, to be discussed in the chapters following, have significantly lower natural frequencies.)

Where the air spring is placed in the suspension makes a difference in the *wheel frequency* on which the vehicle rides. If an air spring is mounted directly above the axle, the ride frequency will be approximately that of the spring values cited above. If, however, the spring is on a linkage, e.g., on a trailing arm suspension, the natural frequency at the wheel will change, depending on the ratio of the pivot-to-spring distance to the pivot-to-axle distance. The natural frequency of the suspension will be proportionately higher than the value of the spring if, on a trailing arm suspension, for example, it is placed behind the axle; and the natural frequency of the suspension will be lower if the spring is between the pivot and the axle.

What the natural frequency for a suspension should be in order to yield a "comfortable" ride has changed with time. In 1936, air spring pioneer Roy Brown observed that, "Each year the [natural] frequency considered most comfortable has been lowered. Some years ago, when the body frequencies [of passenger cars] resulting from the suspensions then used were in the neighborhood of 140 cycles per minute, a frequency of 126 was thought to be ideal. At present, some of the lowest rate suspensions result in body frequencies of from 90 to 110 cycles per minute, while 60 to 80 cycles per minute is considered ideal."[4]

In the decade following Brown's SAE technical paper, most popular passenger cars had steel spring suspensions producing natural frequencies of the order of 90 cpm, although luxury cars of the Cadillac and Rolls Royce class were much lower—in the 50 cpm range.

Even so, considering the small percentage difference in weight of the payload compared to the curb weight of passenger cars equipped with these stiff suspensions, it is not surprising that Detroit engineers saw little advantage in Brown's complex, adjustable-rate air suspension, since the ride really didn't change that much with the steel spring suspensions of the time.

As a result of pioneering studies by British automotive engineer, Maurice Olley (see Chapter 5), as ride frequencies were lowered, it became apparent that vehicle performance and ride comfort, especially for rear-seat passengers, would be improved if the natural frequency of the *front* suspension was lower than that of the *rear*. Partly, this prevented resonant seesawing as the vehicle passed over bumps. On luxury vehicles, natural frequencies as low as 48 on the front and 54 on the rear became common—made possible by the introduction of independent front suspensions.[5]

The 1953 General Motors' luxury coaches with air suspension had a natural frequency of 80 cpm on the rear, and about 5 cpm lower on the front suspension. As first released, this wasn't enough difference. When the new buses were placed into service between Detroit and the Willow Run Airport, the infamous Detroit-Willow Run Expressway became a proving ground for GMC's bus engineers. The concrete highway had been poured during World War II when steel reinforcement was unavailable, and it had uniformly spaced tar strips covering the uneven junctures between the concrete slabs.

The resulting excitation—the very regular input every hundred feet into these 40-foot vehicles with their very soft suspensions (having up to nine inches of available vertical travel)—resulted in unacceptable seesawing at precisely the speeds these buses normally traveled when bringing people into Detroit from the airport. After empirical tinkering with the shock absorbers in cooperation with engineers from GM's Delco Division, the performance was eventually optimized by providing substantial restriction on the rebound stroke of the shocks, but almost none on the jounce stroke.[6]

As noted, even with the spacious "air beam" on early General Motors buses, the natural frequency of its two-convolution pneumatic springs was only depressed to the 75–80 cpm range, and calculations revealed that it was impossible to reduce this value significantly with air springs of Brown's convoluted design, no matter how much larger the displacement volume was made because of the shape of the effective area curve. While these ride frequency values had represented a substantial improvement over prior state of the art for bus suspensions (especially under lightly-loaded conditions), they had become inconsistent with GMC's goals for its "New Look" coaches scheduled for 1960 production release.

This design decision was influenced in part by changes in passenger car suspensions in the meantime. By the mid-1950s, some car manufacturers produced production models whose *steel* spring suspensions had natural frequencies significantly lower than these values. Moreover, both Ford Motor Co. and General Motors had plans to release air suspension on some or all of their 1959 passenger car models, plunging ride frequencies to the 40s and 50s, even on lower-priced cars. In the Detroit automotive market, a strong need had developed for some radical new concepts in air spring design. The next chapter details what was done about it.

Chapter Five

MUTATIONS

At the Annual Meeting of the Society of Automotive Engineers (SAE) in January, 1958, no less than five technical papers introducing new passenger car air suspension systems were presented in the ornate ballrooms of Detroit's Book Cadillac Hotel. A full model year ahead of their original schedules, pneumatic suspension was being announced as a regular production option by Ford, Chevrolet, Buick, Oldsmobile, and American Motors' Rambler. Although they did not present papers at the time, Pontiac also offered air springs in 1958, and Cadillac had put them on one model a year earlier.

Each of these systems included a belt-driven compressor under the hood, a high-pressure air storage tank, leveling valves, copper or nylon air lines, and an assortment of new air spring designs which have become the since-published components of SAE J511A. Illustrated in Fig. 5.1,[1] these flexible configurations were recognizable at the time as, **A**, the GM Technical Center's restricted reversible diaphragm design made by the United States Rubber Co., which was adopted by Buick and Oldsmobile; **B**, the unrestricted reversible diaphragm or "1-X" design developed by Firestone in cooperation with Chevrolet (also used by Pontiac and later by Cadillac); **C**, the Ford reversible rolling sleeve (also used on the Mercury and Edsel) with an external girdle band, developed in cooperation with Firestone and General Tire Co., but produced exclusively by General; and **D**, the Goodyear "rolling lobe" which would have been on the Lincoln if air suspension had not been canceled that year—but which did make its first appearance in 1958 on the rear suspension of the American Motors' Rambler as a small-diameter elongated sausage shape surrounded by a coil spring. This innovative pneumatic spring configuration, which maintained its cylindrical shape without any restrictive metal components, would revolutionize air spring design in the years to follow, as related in Chapter 7.

All of these design mutations had as their primary objective achieving a dramatically low natural frequency in the 40–60 cpm range without the cost and space problems associated with the bulky auxiliary displacement reservoir that Roy Brown's demonstration cars had required (or the massive air beam on the GMC coaches). To this end, the effective area and load-deflection curves of the springs would have to be radically modified from the two-convolution characteristics presented earlier in Fig. 3.3.

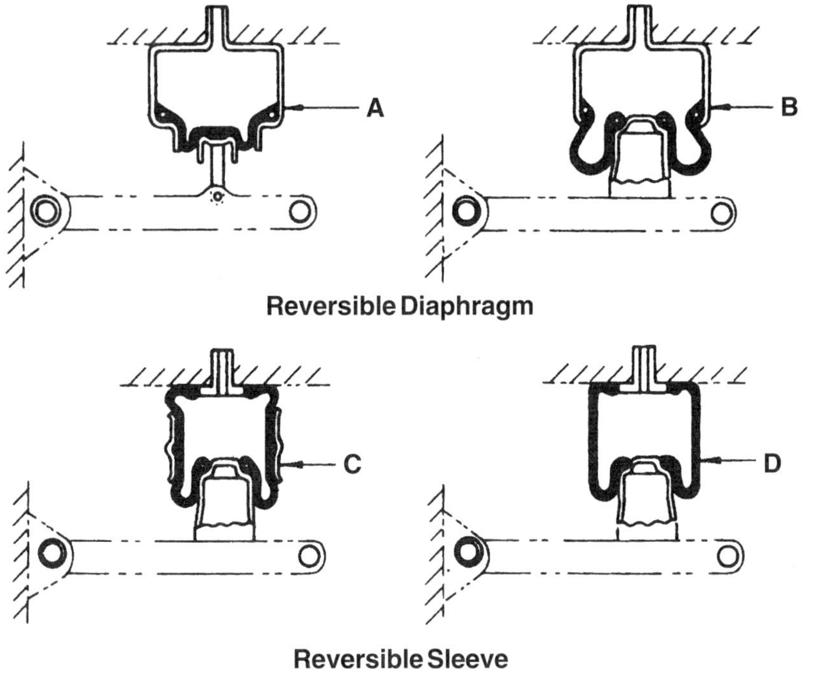

*Fig. 5.1 SAE J511A illustrates the variety of passenger car air spring configurations that emerged in the 1950s. They are recognizable as, **A**, the GM Technical Center's restricted reversible diaphragm made at the time by the United States Rubber Co. and released on the Buick and Oldsmobile; **B**, the unrestricted reversible diaphragm or "1-X" design developed by Firestone for Chevrolet, Pontiac, and Cadillac; **C**, the Ford reversible rolling sleeve with an external girdle band, developed by Firestone and General Tire Co.; and **D**, the Goodyear "rolling lobe" which appeared on the American Motors' Rambler, and would have been on the Lincoln if air suspension had not been canceled that year.*

All of the new designs would have a flexible diaphragm of fabric-reinforced rubber, a contoured metal piston, and sometimes other means to control the effective area during the stroke of the air spring as the flexible part changed shape. Some would bear a remarkable resemblance to designs conceived a century earlier.

Less than a year after successfully launching air suspension on its luxury coaches and transit buses, General Motors' Engineering Staff began working on pneumatic suspensions for passenger cars at the GM Technical Center in Warren, Michigan. There, an advanced Structures and Suspensions group, headed by Von D. Polhemus (a group that included Ming Chi Yew and Max Ruegg) began working on the wide-brimmed "reversible diaphragm" concept (the one designated as **A** in Fig. 5.1), and an attendant pneumatic system.

After an experimental Buick undergoing tests at the GM Tech Center in 1955 won the enthusiastic endorsement of General Motors president Harold Curtis, the system was released for production on the 1957 Cadillac Eldorado Brougham—which became Detroit's first production passenger car with air suspension. The Brougham was a limited-production luxury car with a host of other state-of-the-art offerings, including a burnished stainless steel top, electrically heated seats, and hydraulically cooled brakes. With its selling price in the $20,000 range in 1957, and a cost rumored to be even higher, the car had a limited market, but it was a prestigious offering for Cadillac.

Details of the 1957 Brougham's air suspension were revealed in an SAE technical paper delivered jointly by V. D. Polhemus and L. J. Kehoe Jr., of General Motors Engineering Staff (General Motors Technical Center), and also by F. H. Cowin and S. L. Milliken of Cadillac at the 1957 National Passenger Car, Body and Materials Meeting in Detroit. The front air spring assembly (Fig. 5.2)[2] bears a striking resemblance to J. Hofmann's 1913 Pneumatic Spring patent presented in Chapter 2 (Fig. 2.6). The bulkiness of the resulting package, however, as seen on the front suspension layout (Fig. 5.3)[2] led Cadillac to choose a Type **B** (Firestone "1-X") air spring configuration for its volume production cars.

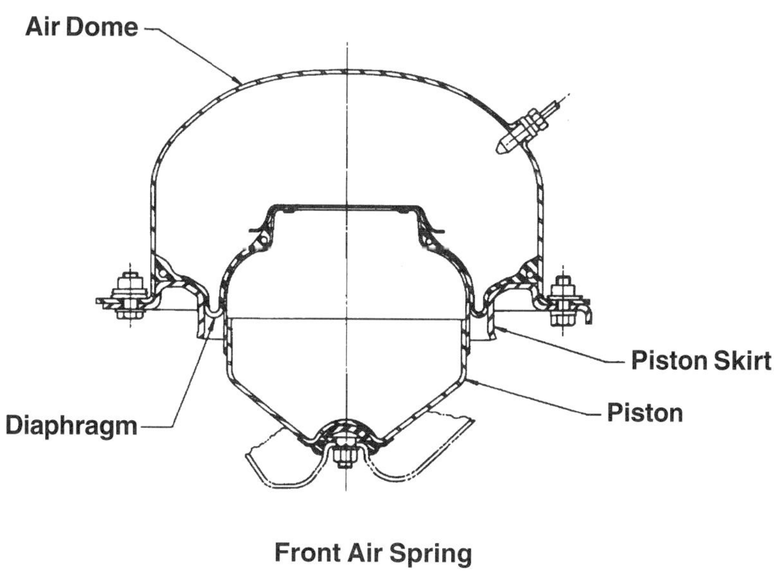

*Fig. 5.2 The Type **A** wide-brimmed "reversible diaphragm" as released on the 1957 Cadillac Brougham. The configuration was the product of a General Motors Technical Center team that included Ming Yew, Max Ruegg, and Von Polhemus.*

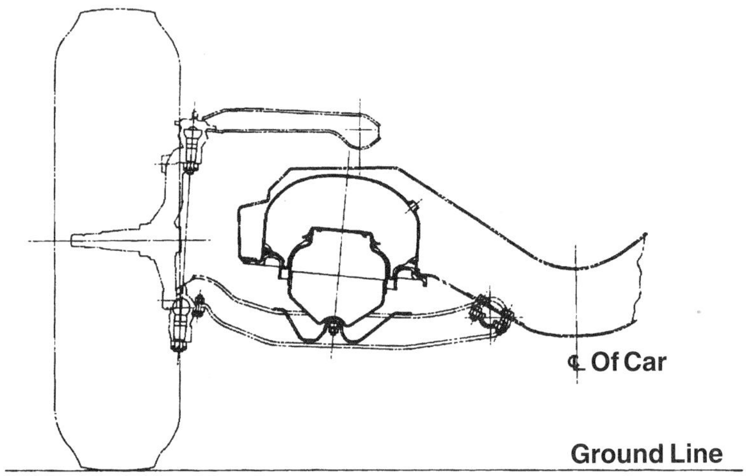

Front Suspension

*Fig. 5.3 Front suspension of the 1957 Cadillac Brougham. The size of the package led Cadillac engineers to opt for the smaller-diameter Firestone 1-X, or Type **B** design for 1958.*

The Brougham had a small, electrically driven compressor mounted on the generator. There were two height-control valves in the rear and one in front. Finding a reliable valve appropriate for a production car proved to be a challenge. Everything on the market was either "too large, too heavy, or too complicated." GM Tech Center and Cadillac engineers ultimately developed a customized unit with two custom Schraeder-type valves (one for intake, one for exhaust), operated directly by the linkage of the lower control arms. The Brougham technical paper goes on to detail a rather complex electrically-operated air distribution system which seems to have anticipated almost any contingency:

"Four solenoid-controlled valves ... meter the flow of high-pressure air to the leveling valves at the wheels. The first pair of valves allow passage of air only when a car door is open or the ignition is on. Of this pair, one valve controls the inlet air to leveling valves, with the other valve controlling the outlet air. When the car doors are closed or the ignition is off, there is no passage of air, and the leveling system is locked out. This provides for all conditions of jacking, shipping and parking.

"The second pair of solenoids control the flow of air under operating conditions. When any door is opened, there is an unrestricted flow of air to the leveling valves, and any change in car height, due to the entrance or egress of passengers, is corrected quickly. This occurs with the ignition on or off.

"During all driving, there is a restricted flow of air through a coined orifice in the solenoid valve. This permits constant adjustment of car height during running to compensate for loss of weight (as with the consumption of fuel), or for added weight (with an accumulation of mud or ice) and it allows a correction in height when the car is driven away after having been parked on a side hill, or with one wheel on a curb or in a deep hole."[3]

A year later, the GM Tech Center system with its Type **A** reversible diaphragm air springs was offered as an option on 1958 Buick and Oldsmobile models. The Buick and Oldsmobile systems were very similar. Fig. 5.4, reproduced from Ralph W. Perkins's 1958 Oldsmobile paper,[4] shows a cross section of the air spring assembly, with its contoured metal piston, a formed metal air chamber into which the piston could displace the diaphragm, and a concentric sheet metal extension from the bottom of the chamber

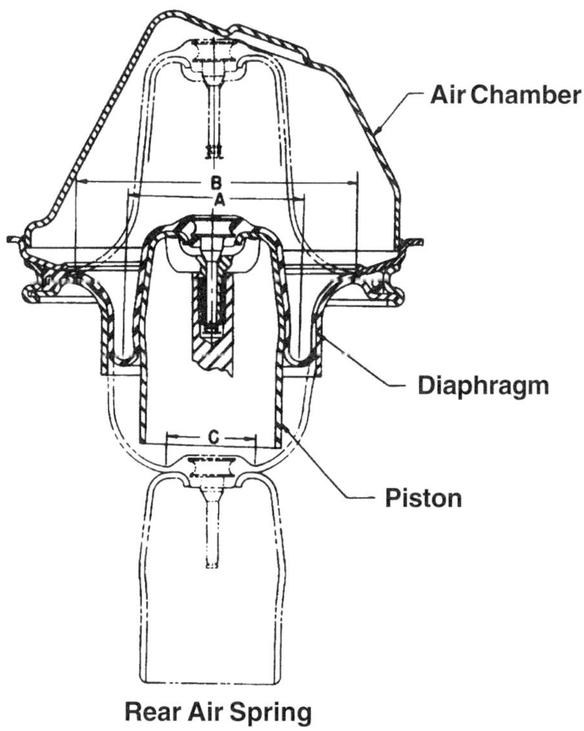

*Fig. 5.4 Type **A** restricted reversible-diaphragm air spring as released on the rear suspension of the 1958 Oldsmobile.*

which mechanically guided the effective diameter of the spring through the curved path illustrated in Fig. 5.5 (from the Buick paper by Forest McFarland, Gail Peckham, and Eric Dietrich).[5] This S-shaped effective area path resulted in the load-deflection curve shown in the Oldsmobile paper (Fig. 5.6.)

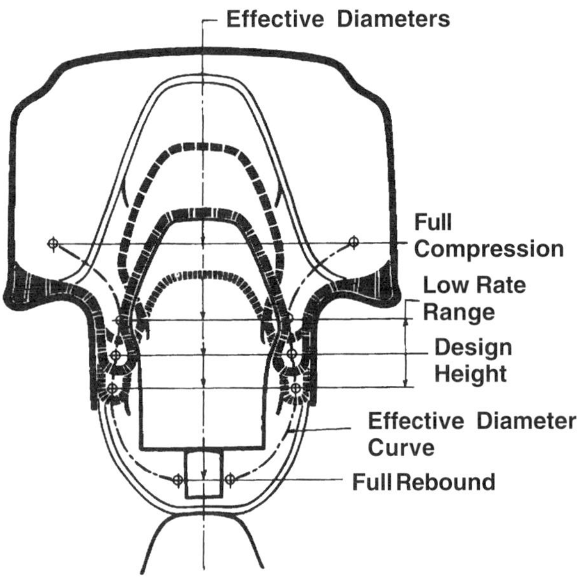

Fig. 5.5 Showing how the effective diameter (radius center of the rolling meniscus) of the Buick reversible diaphragm, guided by the contoured piston and external metal skirt, tailors the change in effective area with stroke in order to optimize the load-deflection curve of the spring; from the SAE paper by McFarland, Peckham, and Dietrich.

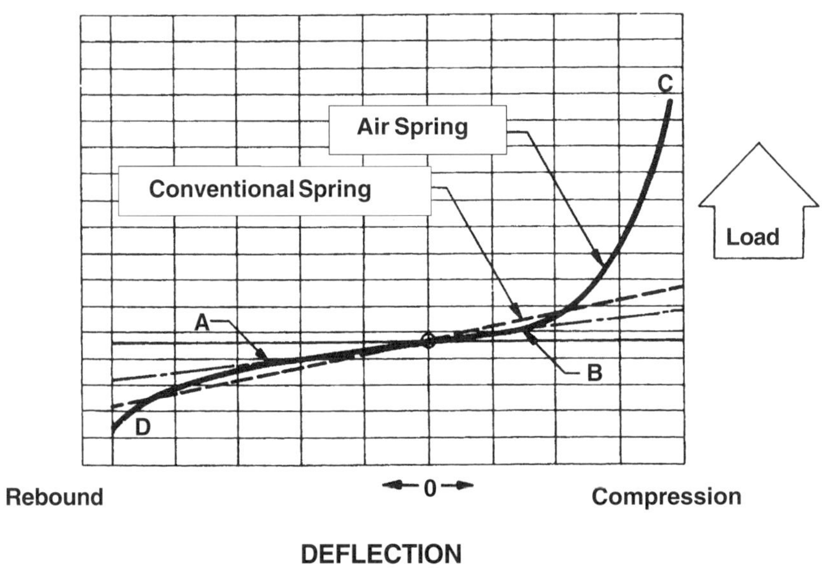

*Fig. 5.6 The resulting load-deflection curve of the Type **A** spring has a flat, low-rate portion on either side of the design position of the spring, with the curve turning sharply upward as the suspension goes into jounce, and turning downward as it goes into rebound.*

Typical of most load-deflection curves produced by this new generation of passenger car air springs, the curve has a nearly flat, low-rate gentle slope for an inch or two [25–50 cm] on either side of the design position of the spring, with the curve turning sharply upward near the end of jounce travel, and with a similar downward-turning portion as it goes into rebound. This design feature limits over-travel and "crash-through" into the suspension bumpers when the soft suspension is driven over rough roads.

Listed in Perkins's Oldsmobile paper are the components of the pneumatic system, layed out graphically in Fig. 5.7:[4]

- Four-link rear suspension
- Diaphragm-type air springs
- Height control valves
- Air compressor, mounted integrally with the power steering pump
- Oil and moisture separator
- High-pressure tank of extra-large volume
- Low-pressure tank
- Pressure regulator and elevator valve unit
- Copper air lines throughout the system

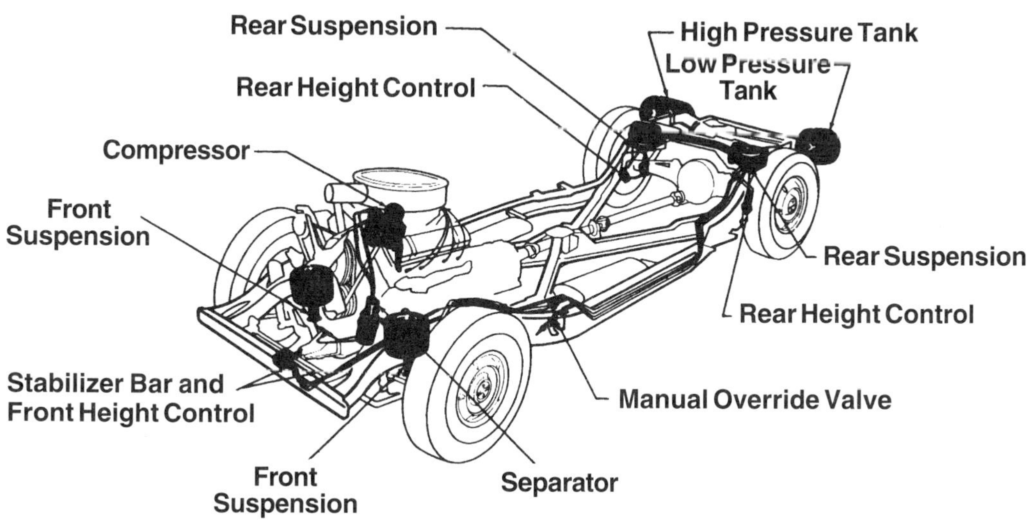

Fig. 5.7 Components of the 1958 Oldsmobile air suspension system as documented in the SAE paper by Ralph W. Perkins.

In his paper, Perkins goes on to explain that, "A four-link suspension is used at the rear to hold the rear axle in position, a function normally performed by rear leaf springs. The rear air springs are mounted on the axle housing independent of the suspension links, and are, in this manner, used solely to control ride. The upper suspension links are diagonally mounted to provide lateral control of the axle, and together with the lower links, absorb drive and braking torque and control fore-and-aft motions. The lower links are adjustable to permit optimum setting of the differential pinion angle. The four-link suspension provides a high rear roll center, and is designed to minimize rear 'squat' and brake lift."

The Buick suspension provided a softer ride than that of the Oldsmobile. And because Buick had a torque tube drive, the linkage geometry of the suspension differed somewhat from that of the Oldsmobile; but these details are not provided in the Buick paper.

The compressors on both cars, driven by a V-belt at slightly above crankshaft speed, had a stall pressure of 290 psi [20 bar]. As the Buick paper explains, "Air from the compressor passed through a check valve to a high pressure storage tank of 820 cubic inches [13.4 liters] volume, located at the front of the chassis. From the storage tank, the air is conducted to a reducing valve in the manual override valve set at 145 psi [10 bar] to maintain constant pressure in the supply system. This constant pressure controls the rate of response of the air springs.

"From the manual override valve, air is piped to a front height control valve through a check valve. The height control valve is operated by the lever from the center of the stabilizer bar to obtain the average of the car height at both front wheels. This permits air to flow to the air springs when the car is below design height or to escape from the air springs when the car is above it.

"The air from the front height control valve to the air springs passes through a cross-flow valve, which consists of a double check valve permitting unrestricted flow to either, but restricted flow from each to minimize cross-flow from one air spring to the other under roll [tilting sideways when rounding a curve].

"Air flows to the rear right and left height control valves, controlling the height of the car at the right and left air springs in the same manner as the front, except that each wheel is controlled separately."[5]

The Buick authors explain that *four* valves had been tried experimentally, one at each wheel. But this resulted in instability and "hunting." Since three points determine a plane, four valves, it was learned, "fight with one another," and three were, therefore, released for production.

The plumbing for the pneumatics was a semi-closed system, returning air exhausted from the height-control valves to the compressor inlet or to the air cleaner, "depending on whether or not the compressor is running. Any make-up air required by the compressor ... is drawn through the air cleaner connection." The storage tank was also "used as a moisture accumulator. Any casual water collected is drained on the lubricating schedule."[5]

Both the Buick and Oldsmobile air springs operated at 100 psi [6.9 bar] when the vehicle was fully laden. As indicated above, each of these vehicles had a manual override valve, as did the pneumatic suspensions of some other General Motors vehicles—a valve complex that not only provided a means to adjust the pressure regulator, but, as the Buick paper explains, it "has another function—that of raising the car over four inches when a knob under the instrument board is pulled. This operation causes a valve plunger in the override valve to shut off the exhaust line to the compressor and open the exhaust lines of the height control valves."

When actuated by the driver, this feature admitted 145 psi [10 bar] air directly to the springs, overruling the height-control valves, and causing the air springs to lift the car up against the rebound bumpers—a rise of over four inches [10 cm] in front and approximately five and a half [14 cm] on the rear.[5]

"This function is useful for abnormal conditions," the paper continues, "such as pulling through mud holes, deep snow, high crowned roads, or loading the car on unusual lift hoists. While the ride is, of course not comparable to normal, this function can be very helpful. While driving over one of the trails in the Rocky Mountains during development, our drivers used this device to get through one bad spot. A tourist, seeing our cars go through, attempted to follow and ended up stuck in the middle of the spot. After pulling him out, our people left him wondering how we were able to get through so easily."[5]

Buick offered air suspension as an option on all models at an extra cost of $188. Oldsmobile offered it on all of its models, except those that were not equipped with power steering, at a premium of $215. Although both of these GM cars both used **A**-type air springs manufactured by U.S. Rubber, their performance characteristics differed somewhat. Buick, which traditionally had always favored a very soft ride, had a significantly lower natural frequency on the front, as indicated in Table I. To this end, the Buick suspension used a smaller diameter piston and had a larger air spring volume. The Buick, it will be noted, also provided substantially more stroke.

TABLE I

A-TYPE PNEUMATIC SPRINGS

Property	BUICK Front	BUICK Rear	OLDSMOBILE Front	OLDSMOBILE Rear
Natural Frequency	43	54	46	54
Suspension Travel				
Jounce	4 in. (10.2 cm)	4.2 in. (10.7 cm)	2.3 in. (5.8 cm)	3.8 in. (9.7 cm)
Rebound	4 in. (10.2 cm)	5.3 in. (13.5 cm)	2.4 in. (6.1 cm)	5 in. (12.7 cm)
Design Pressure	100 psi (6.9 bar)	100 psi (6.9 bar)	100 psi (6.9 bar)	100 psi (6.9 bar)
Air Spring Volume	300 in.^{3} (1.94 L)	235 in.^{3} (1.5 L)	295 in.^{3} (1.9 L)	240 in.^{3} (1.55 L)

Type **B**, the second air spring design configuration in Fig. 5.1, was the product of a joint development between General Motors' Chevrolet Division and Firestone Industrial Products Company. Chevrolet's chief engineer, Ed Cole, familiar with Firestone's work at GM Truck & Coach Division, had asked for the company's help in developing a system for his division.

The first meeting with Chevrolet, held at the General Motors Building on Detroit's Grand Boulevard, was attended by Firestone engineers Fred Haushalter, Roy Brown, and Herbert H. Deist, then assistant manager of Firestone's Engineering Laboratory in Akron (the department responsible for both equipment and process development for the company).

Their host was Morris Olley, distinguished British automotive engineer who had originally come to the United States to be chief engineer of Rolls Royce USA in Springfield, Massachusetts. Making major design and manufacturing changes there, Olley is credited with building the world's finest *Silver Ghost* Rolls. Unlike its British cousin, the American Rolls had a light, rugged Brewster body, softer springs which produced a better ride, but had better handling than the parent company's car. In the 1920s and '30s it had been Olley who pioneered the concept that, to improve the ride for rear-seat passengers, the natural frequency of the front suspension had to be substantially decreased (from then-common 90 cpm to, ideally, 50–60 cpm—an impossible goal with the leaf springs of the time). When Rolls closed the Springfield plant, Olley was hired by E. Seaholm, Cadillac's chief engineer to improve the ride of GM's flagship car. He had since become a General Motors staff consultant serving several divisions of the company.[6] Olley would play a major role in the development of the Chevrolet suspension.

During their first meeting, the Firestone group was invited to have lunch in General Motors' executive dining room with Chevrolet's chief engineer, Ed Cole (who would later become general manager of the Division). When Cole asked Roy Brown to explain how his product worked, Brown's condescending, simplistic discourse and generally arrogant demeanor was less than endearing to his hosts. It would be his last trip to Detroit. Future Firestone contributions to the Chevrolet development effort would come from Fred Haushalter and Herb Deist (who is quick to defer credit to his innovative project engineer at the time, Jack Hollis).

During the next several years, Haushalter and Deist would also call on the engineering departments of Ford Motor Company, American Motors, and Packard, as well as other General Motors divisions.

Much more than an administrative program manager, Morris Olley literally rolled up his sleeves and personally worked out all of the mathematics governing the placement of the air springs and related suspension components on GM's new "cruciform" frame (which would also accommodate Chevrolet's new coil spring rear suspension). Olley's contributions would include establishing the centers of rotation for the upper and lower control arms, geometry of the steering mechanism, and the caster and camber of the front wheels, with particular attention to such sophisticated parameters for the time as oversteer/understeer characteristics, anti-dive upon braking, and rear-axle steer—all this at a time when engineers' calculating tools were limited to a slide rule and a Friden calculator.[7] Olley was assisted in his analysis by Chevrolet engineer Robert Schilling and the authors of the SAE technical paper which is quoted below.[8]

Fig. 5.8[9] shows a 1956 Firestone layout of a front suspension under consideration at the time, showing the placement of a radically new (Type **B**) air spring design, shown in cross section in Fig. 5.9.[10]

The rolling diaphragm configuration was conceived by Jack L. Hollis. Hollis also designed the building equipment for the Type **B** air spring. He would later build Firestone's Decatur tire plant and would be appointed chief engineer of more than one rubber company before ultimately becoming president of the Carlisle Tire & Rubber Company.

Hollis explains that his design represented a simplification of the more complex "2-X" air spring, a configuration first conceived in 1916 by Herbert F. Rudd—a design that had been under consideration for passenger cars in Roy Brown's department for some time.

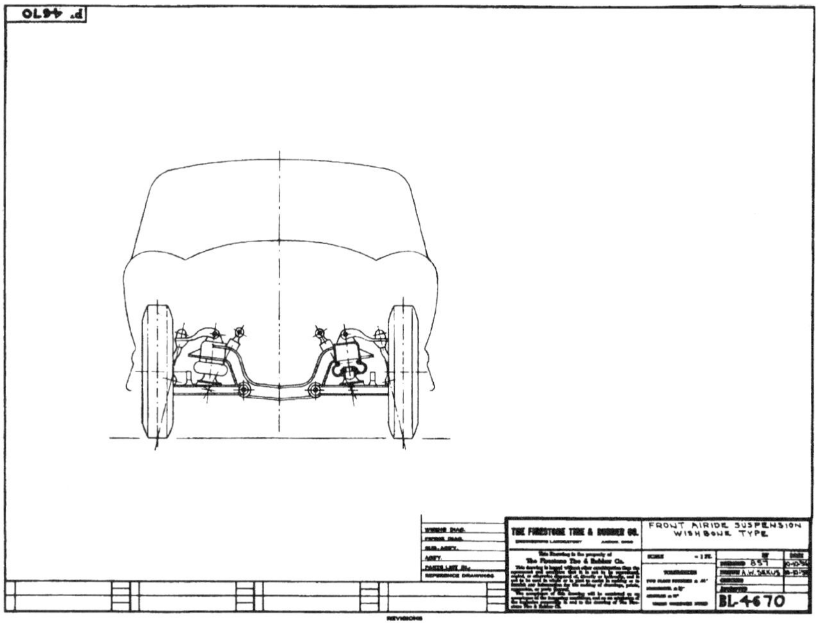

*Fig. 5.8 A 1956 Firestone proposal drawing showing the placement of the Type **B** air spring, a design conceived by Firestone engineer Jack Hollis.*

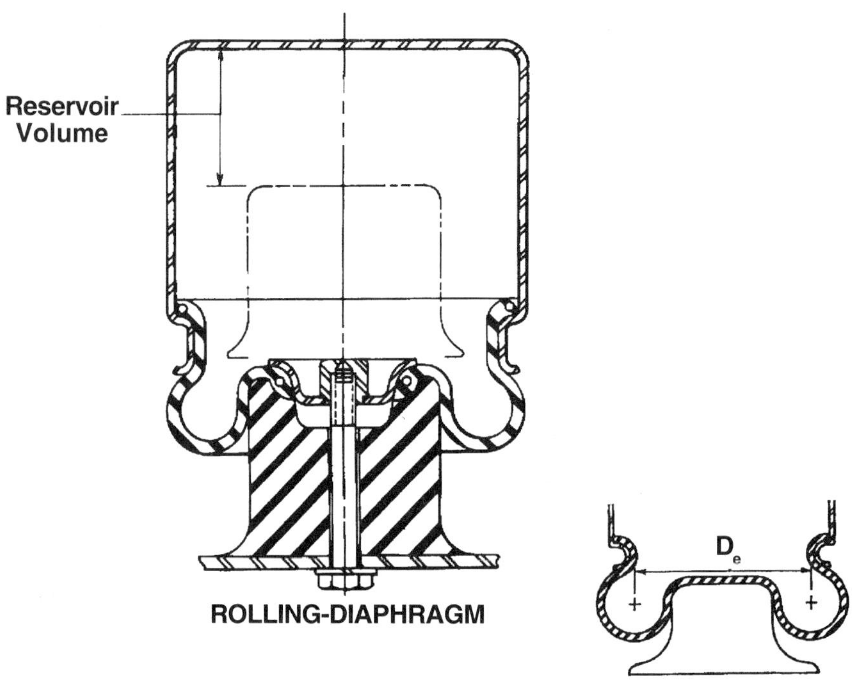

*Fig. 5.9 Detail of the Firestone Type **B** "1-X" design, showing the effective diameter of the rolling diaphragm at design position (lower right).*

(The 2-X design was later offered to Ford Motor Company—avoiding disclosure of the Type **B**, which had been developed for Chevrolet.) When Brown saw Hollis's design, he was unimpressed, and he scolded Hollis for wasting company time. But when a Firestone patent for the 1-X, Type **B** design was applied for, "for the good of the company" the patent was issued in the names of both Brown and Hollis, and was published as "Brown et al."[11]

The new configuration became known as the "1-X" series. The "1" stood for "single convolution" and "X" for the secrecy surrounding the project at the time. Substantially smaller in diameter than the air chamber on the GM Technical Center reversible diaphragm on the Buick and Oldsmobile (Type **A** above), the 1-X had a cylindrical canister into which the upper bead of the air spring could be snapped into place behind a retaining flange. The bead had a lip that sealed against the interior of the metal canister somewhat like a tubeless tire seals against the wheel.

Instead of the wide brim of the GM Tech Center diaphragm, the 1-X has the appearance of a Firestone single-convolution bellows built with the lower bead much smaller than the top one, so that one bead can pass through the other, favorably manipulating the effective area characteristic in the process. Sealing the small end is an adhered, molded-in metal cup which engages the top of the piston.

Recognizing that the path of the effective area loci would also be a function of piston shape, a study was undertaken to determine the effects of straight, tapered, and "hourglass" pistons on the effective area vs. stroke curve. Results are documented in Fig. 5.10.[9]

A typical effective area characteristic for a production 1-X spring, together with the resulting load-deflection curve, is shown in Fig. 5.11.[10] The S-shaped effective area curve *and* the resulting load-deflection curve are quite similar to that of the GM Tech Center diaphragm of the Buick and Oldsmobile suspensions, although the increase in rate at the end of the jounce stroke of the 1-X is achieved by flaring the piston at the base, spreading out the effective diameter locus points—rather than by permitting these radius centers to expand outward as the center of the diaphragm enters the dome-shaped displacement chamber on the GM Technical Center design.

Significantly, the slope of the center section of the effective area curve is actually *negative* in the center section near the design position. Although this feature reduces the volume required to achieve a low dynamic (adiabatic) spring rate, it would later produce totally unforeseen phenomena on experimental test cars when the static (essentially isothermal) spring rate became negative.

This sophisticated tailoring of the effective area curves for the new generation of passenger car air springs complicated the lives of process development engineers in Firestone's Engineering Laboratory, who inherited the responsibility for devising ways to manufacture these products. Unlike two-convolution air springs, which could be layed

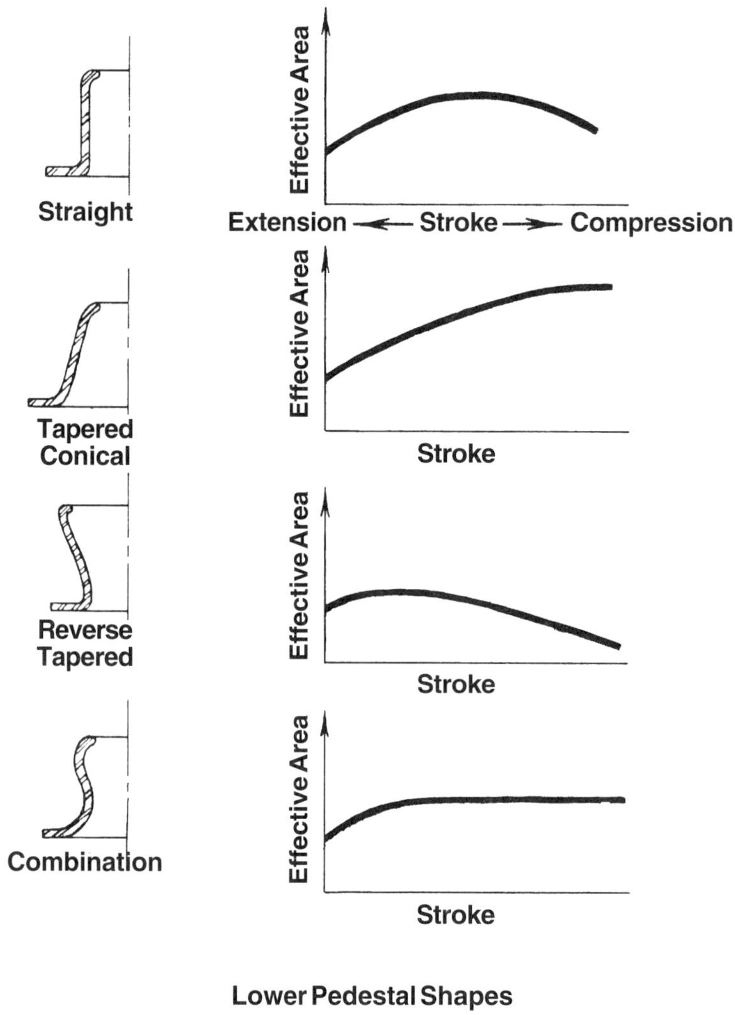

*Fig. 5.10 Results of a study by Herbert Deist's (Firestone) Engineering Laboratory, showing the effect of the shape of the piston on the load-deflection curve of a Type **B** air spring.*

up on a cylindrical building drum (see the relatively simple mold "prep" shown earlier in Fig. 2.10), provision had to be made for expanding the upper beads of the new designs to more than twice the diameter of the lower beads. In designing the building equipment to produce the Chevrolet 1-X spring on a volume basis, Jack Hollis provided a building drum with mechanical fingers that expanded the end of the cylinder into a cone, as seen in Fig. 5.12.

The resulting conical shape of the mold prep is shown at the lower right of the first page of Tom Bank's "Apparatus and Method" patent filed in 1956, as shown in Fig. 5.13. This figure also pictures an enlarged portion of the Chevrolet layout seen above in Fig. 5.8, and a cross section showing the distortion the flexible diaphragm had to go through in the arcuate motion of the air spring.[12]

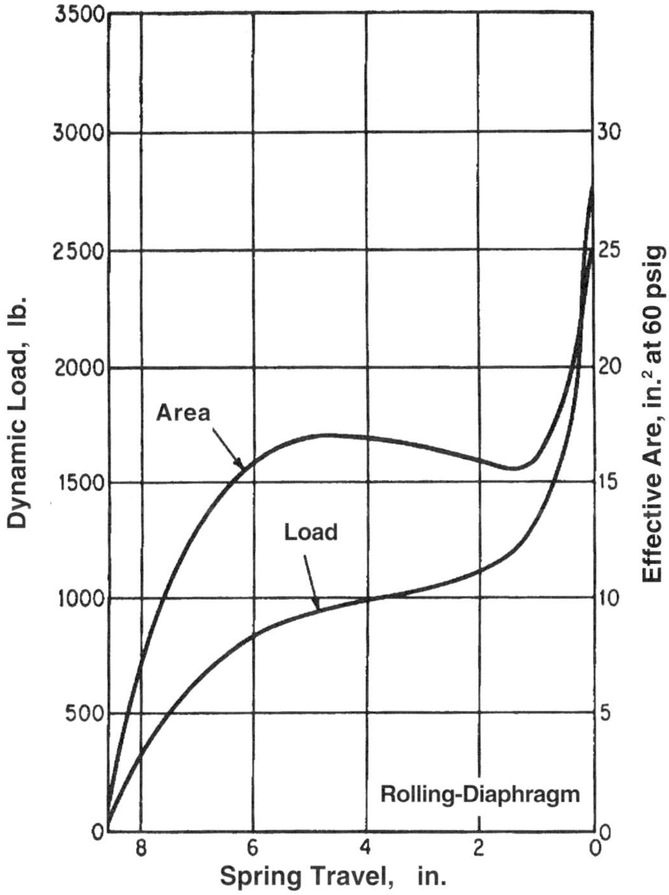

*Fig. 5.11 Typical effective area and load-deflection characteristics for a Type **B** air spring.*

*Fig. 5.12 Automated building machine for Type **B** air springs, as designed by Jack Hollis.*

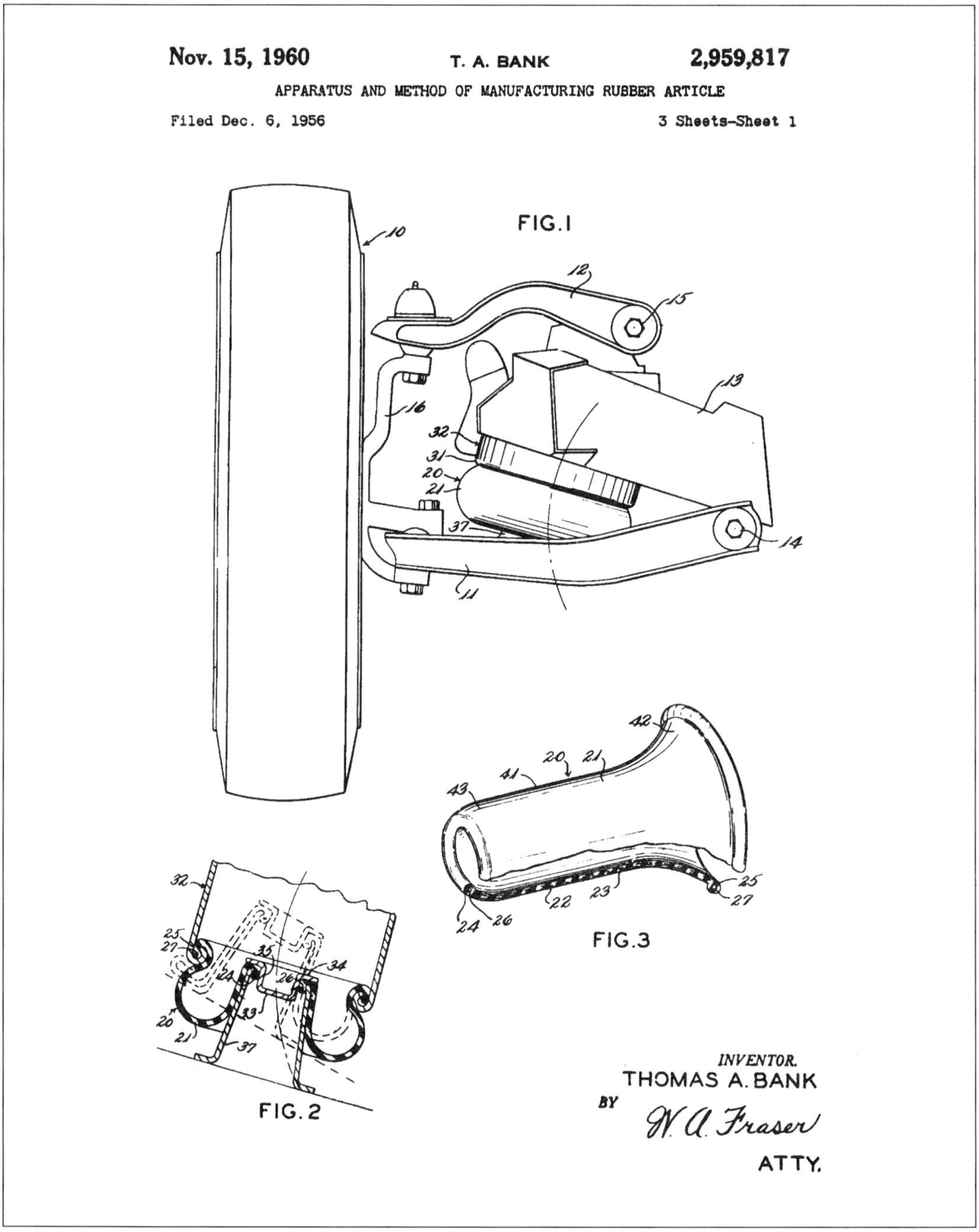

*Fig. 5.13 Megaphone shape of the "moldprep" as built for a Type **B** air spring is seen in the center of this Tom Bank patent, which also illustrates application of the part on the Chevrolet front suspension.*

The mold prep is built like that of a two-convolution bellows, beginning with a "pure gum" (unreinforced) inner liner. Two plies of tire cord are then rolled on, with the cords intersecting at an acute angle; an outside layer of abrasion-resistant cover stock completes the assembly. The cords are turned up and stitched down around a circular bead wire at each end. One of the beads on this design is substantially larger in diameter than the other.

Figs. 5.14 and 5.15, which represent the next two pages of the Bank patent, show how the megaphone-shaped mold prep is placed on a flanged vertical mandrel in the air spring mold, together with the metal insert "end closure" which will be bonded in place to seal the lower bead. The mold prep is inflated with "shaping air," expanding the rubber and reinforcing cord structure, as it fills the hot mold cavity simultaneously with the closing of the mold—after which the internal air is replaced by steam to cure the interior of the part.

Fig. 5.16 pictures two finished molded 1-X (Type **B**) air springs. The upper one, labeled 1X24A, is the Chevrolet air spring.

The 1958 Chevrolet SAE paper by Kai Hansen, J. F. Bertsch, and R. E. Denzer,[13] documents the pneumatic system employing the Type **B** air spring (Fig. 5.17). The Chevrolet system was somewhat simpler than the one employed by Buick and Oldsmobile. As the authors explain, "Four air spring assemblies are used in the Level Air Suspension system. Integral with each of three of the units is a leveling valve which automatically controls the height at all times. Since three points define a plane, only three points may be used to establish the attitude and height of the car. Therefore, two leveling valves are used in the front, one in each spring, and a single valve is installed in the left rear. The left rear spring is cross-connected to the right rear spring by a balance line. In order to keep unrestricted cross flow between the two from interfering with the handling of the vehicle [e.g., roll], a 0.020 in. [0.5 mm] orifice is included in the balance line.

"A single cylinder air compressor [Buick and Oldsmobile used two-cylinder compressors], belt driven from the crankshaft, supplies air to the system, directly charging the high pressure reservoir when the engine is running."

The pressure in this tank was maintained at 225–250 psi [15.5–17.2 bar]; minimum operating pressure for the system was 180 psi [12.4 bar]. The "1X24A" air spring operated at 100 psi [6.9 bar] when the car was fully loaded. Significantly, unlike Buick and Oldsmobile (or the Cadillac Brougham), Chevrolet did not provide a pressure regulator between its air tank and the leveling valves, a decision that would create major

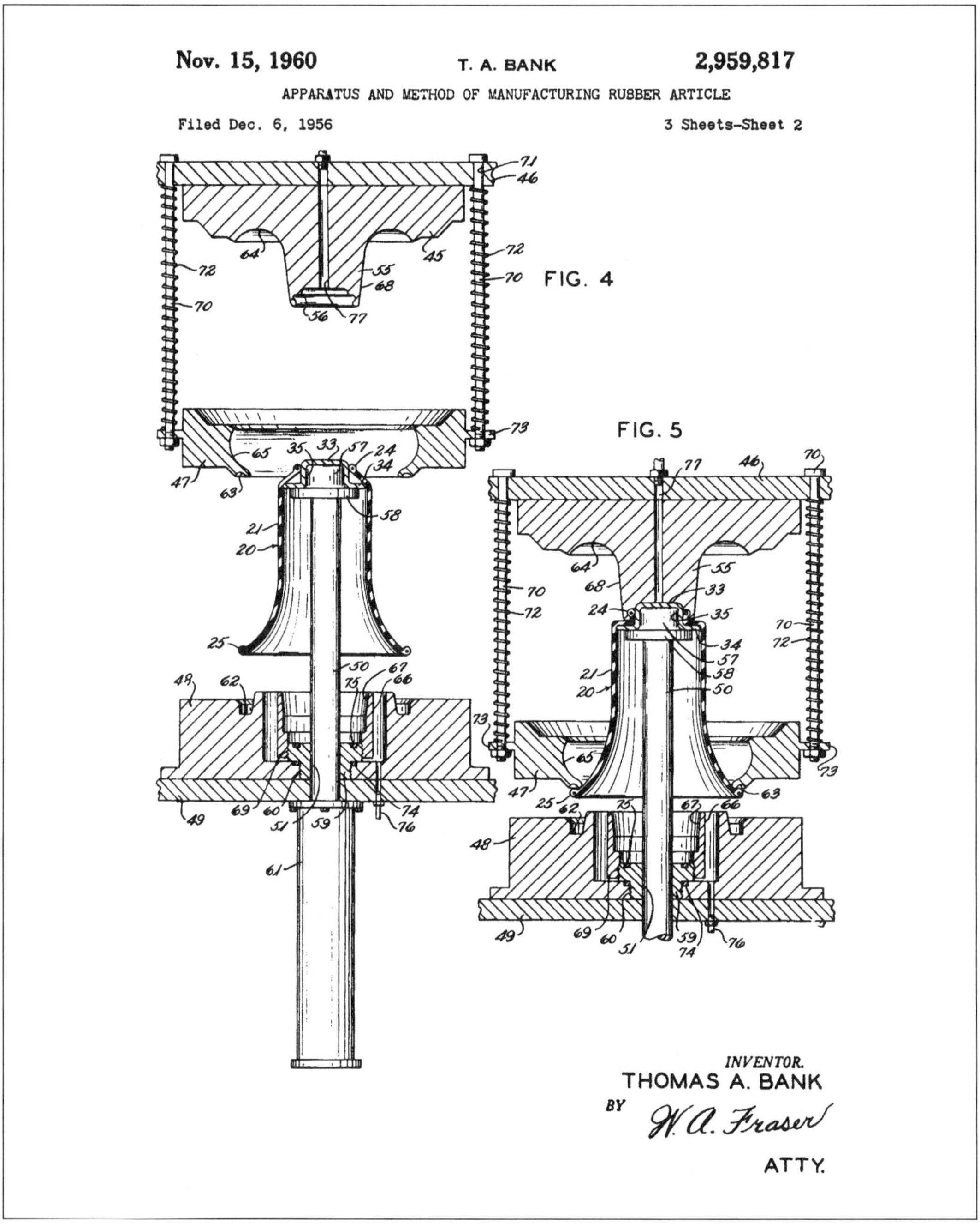

*Fig. 5.14 Further illustrations from the Bank patent show how Firestone's molds shaped the built "prep" into the Type **B** air spring configuration before curing the part.*

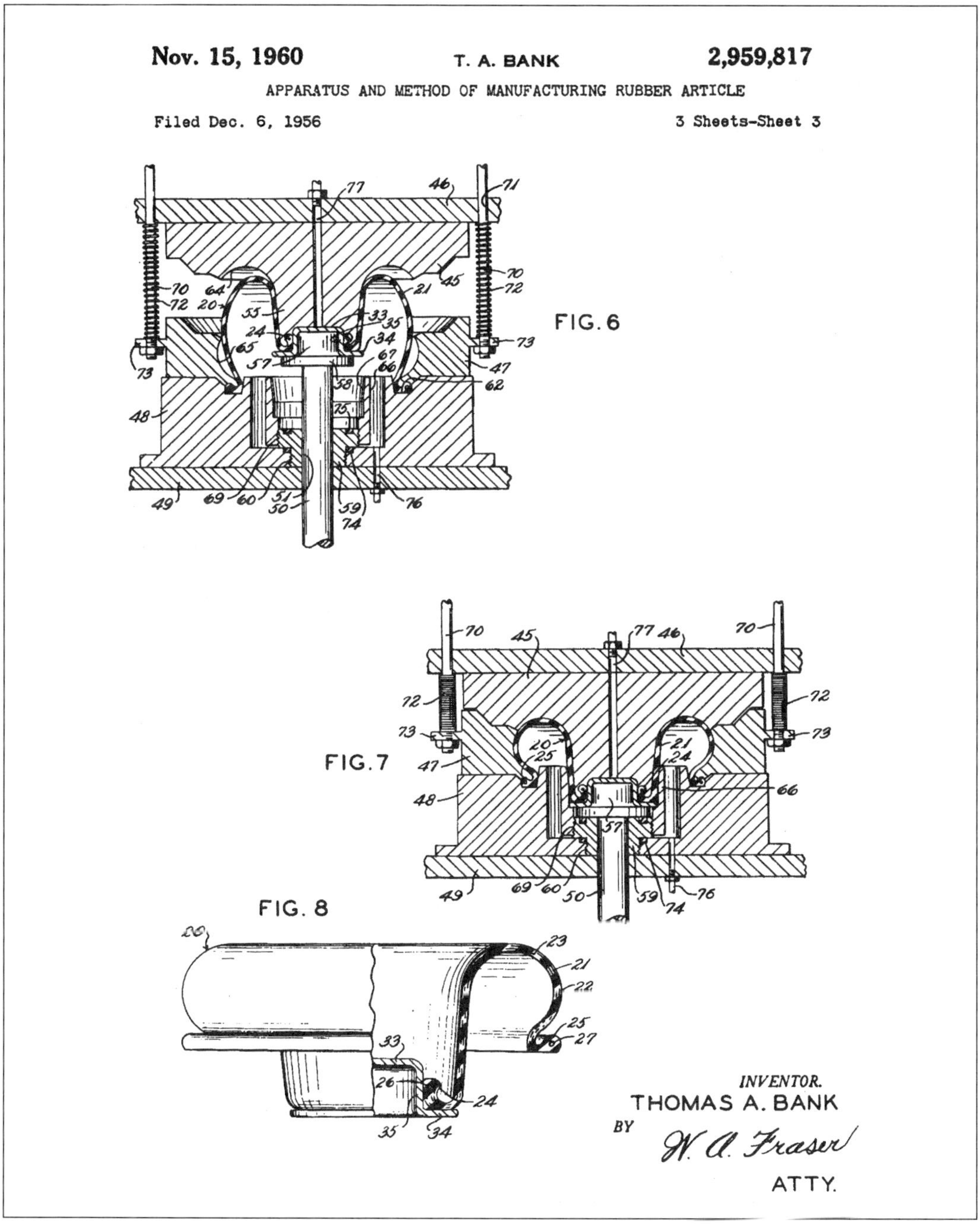

*Fig. 5.15 Further illustrations from the Bank patent show how Firestone's molds shaped the built "prep" into the Type **B** air spring configuration before curing the part.*

*Fig. 5.16 Molded Type **B** air springs, having beads of radically different diameter, with the piston end sealed by a molded-in metal cup. The large top bead has a molded lip designed to seal against the inside of the metal reservoir can into which the part will be snapped in place.*

problems in the field after production release—as will be discussed, together with other unanticipated eventualities, in the next chapter.

Like GM Truck & Coach Division, Chevrolet established very high standards for its air springs: "Current practice calls for rejection of any group of bellows where test samples do not satisfactorily exceed 100,000 full stroke cycles on the [laboratory test] machine without failure. ... One of the requirements specified for the air spring assembly is that it must not lose more than 1 psi [0.07 bar] over a 3 day period when inflated to 75 psi [5.1 bar]. The air spring assemblies have been able to satisfy this specification after 35,000 miles of durability testing." [13] Subsequent field experience, unfortunately, would not match these optimistic test car data.

The pneumatic circuitry for the Chevrolet system is shown in Fig. 5.18.[13] It is apparent that Chevrolet engineers gave special attention to the physical handling of cars with the new system during assembly in the factory, as well as when on the road, and even when the engine was *not* running:

"Lines connect to a junction block mounted on the frame. A manual valve in the junction block prevents the exhaust [of] air from the system. This makes it possible to assemble and inflate the entire air suspension prior to installing the body and the front sheet metal which mounts the reservoir."[13]

And further, with regard to performance on the road: "An interesting phenomenon is the action of the suspension under zero reservoir pressure, a condition that may occur with either a belt or compressor failure. The suspension does not become completely deflated, as at first may be expected, but only deflates to a point where the air exhaust valves do not open during vehicle motions. The car then rides slightly low on the air trapped in the bellows."[13]

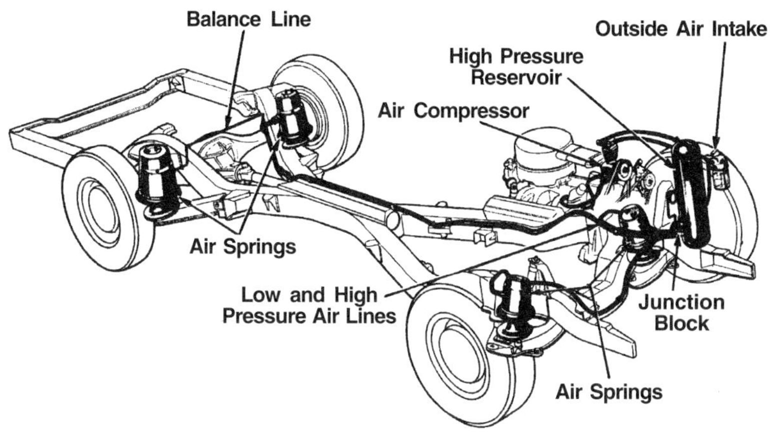

Fig. 5.17 Components of the Chevrolet air suspension system as detailed in the SAE technical paper by Hansen, Bertsch, and Denzer.

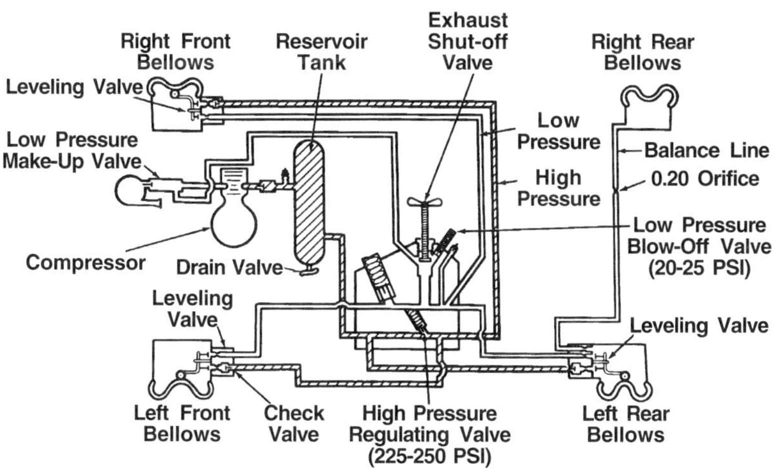

Fig. 5.18 Pneumatic circuitry of the Chevrolet air suspension system.

An innovative departure from that of other GM cars was Chevrolet's placement of the leveling valve *inside* the air spring displacement chamber (Fig. 5.19) to protect it from salt and road grit, permitting the use of a zinc die-cast assembly in which the operating mechanism was partially exposed. As the paper explains, "A spring-loaded lever rides on top of the piston in the air spring. The lever, through linkage, opens and closes the orifices in the valve, as determined by the position for the piston. At design position, both orifices are closed. The metering orifices are incorporated in the valve seats."

The last paragraph of the paper reveals one more Chevrolet innovation—one that would lead to problems, however: "The Trans-Canada Highway, between Port Arthur and Toronto, was one of the roads used for testing of the air suspension system. Temperatures which dropped as low as minus 30° Fahrenheit [–34°C] indicated a need for preventing icing of moisture introduced with make-up air. The addition of a simple alcohol evaporator at the air intake has provided the needed protection."[13]

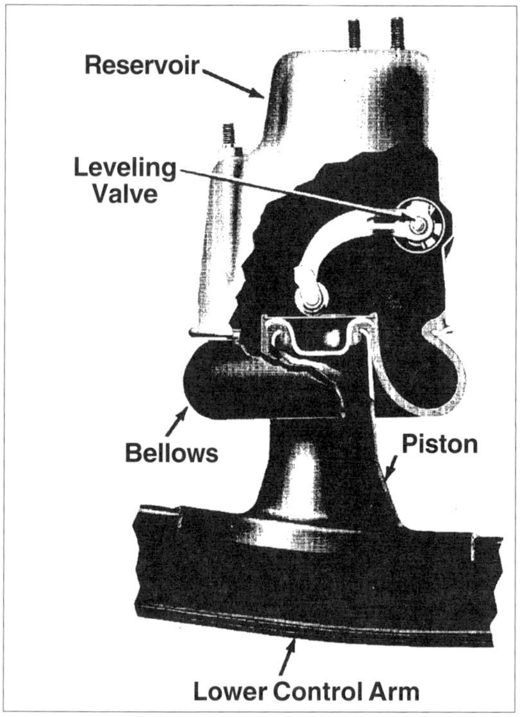

Fig. 5.19 Showing how the height-control valve on the Chevrolet was located inside the air spring to protect it from road salt and grit.

Chevrolet offered air suspension, at a premium of $124, on all of its models (except those cars equipped with fuel injection, perhaps due to difficulties with placement of the compressor).

Both the Pontiac Division of General Motors and Cadillac had begun their air suspension development work with the GM Technical Center Type **A** diaphragm, but both divisions ultimately chose the Firestone 1-X Type **B** configuration, primarily because it was a more compact package, making it easier to integrate into the chassis (reference Figs. 5.2 and 5.3, above).[14] Cadillac delayed its decision until it had completed extensive testing of cars with both spring options, postponing its wider release of air suspension until the 1959 model. Both the Cadillac and Pontiac systems were somewhat more complex than the Chevrolet offering, employing more sophisticated leveling valves, for example, *and* a pressure regulator which reduced high-pressure air to 125 psi (8.6 bar) before delivering it to the height-control valves.

Like Buick and Oldsmobile, Cadillac provided the driver with an "Air Lift" lever to inflate the air springs into full rebound position. However, because the 1-X (**B**-type) spring does not have an external supportive metal skirt as does the GM Tech Center design (or the Bell or Hofmann configurations), this feature would lead to (sometimes explosive!) problems—to be documented, along with other pneumatic adventures, in the next chapter.

Air suspension was offered as an option on all Cadillac models at an extra cost of $215. It was standard equipment on the revised Eldorado Brougham. Pontiac's premium was $175, an option available on all models.

Table II documents the performance differences between the air springs of the three General Motors cars employing the **B**-type pneumatic spring design.

TABLE II

B-TYPE PNEUMATIC SPRINGS

Property	CADILLAC Front	CADILLAC Rear	PONTIAC Front	PONTIAC Rear	CHEVROLET Front	CHEVROLET Rear
Natural Frequency	41	50	52	56	53	64
Suspension Travel						
Jounce	3.9 in. (9.9 cm)	4.3 in. (10.9 cm)	4.5 in. (11.4 cm)	4.1 in. (10.4 cm)	3.9 in. (9.9 cm)	4.3 in. (10.9 cm)
Rebound	4.7 in. (10.2 cm)	5.5 in. (13.5 cm)	5 in. (6.1 cm)	4.8 in (12.7 cm)	4.3 in (10.9 cm)	5.7 in (14.5 cm)
Design Pressure	90 psi (6.2 bar)	90 psi (6.2 bar)	100 psi (6.9 bar)	100 psi (6.9 bar)	90 psi (6.2 bar)	52 psi (3.6 bar)
Air Spring Volume	338 in.^{3} (2.18 L)	221 in.^{3} (5.61 L)	230 in.^{3} (5.84 L)	204 in.^{3} (5.18 L)	n.a. n.a.	n.a. n.a.

Tracing the genesis of the **C**-type of air spring configuration depicted in Fig. 5.1 is complicated. The concept goes back at least to 1913, when Herbert F. Rudd of Chicago filed his patent application for a Pneumatic Cushion Structure. Rudd's design, as seen in Fig. 5.20, resembles two 1-X type springs clamped together by means of a wide girdle band, with the resulting cylindrical middle section also providing a displacement volume between the two facing diaphragms.[15]

A later Firestone design, shown in a (circa 1955) Bank patent drawing (Fig. 5.21), and revealed in the 1957 Deist paper (Fig. 5.22), provided for molding the rubber element as a single piece, as pictured in Fig. 5.23.[9]

In this early configuration, the air spring has two pistons—one on the top and one on the bottom—with a wide external girdle band of steel that restrained what would otherwise be the bulging center section of the rubber part. At Firestone, this configuration came to be known as the "2-X" design.

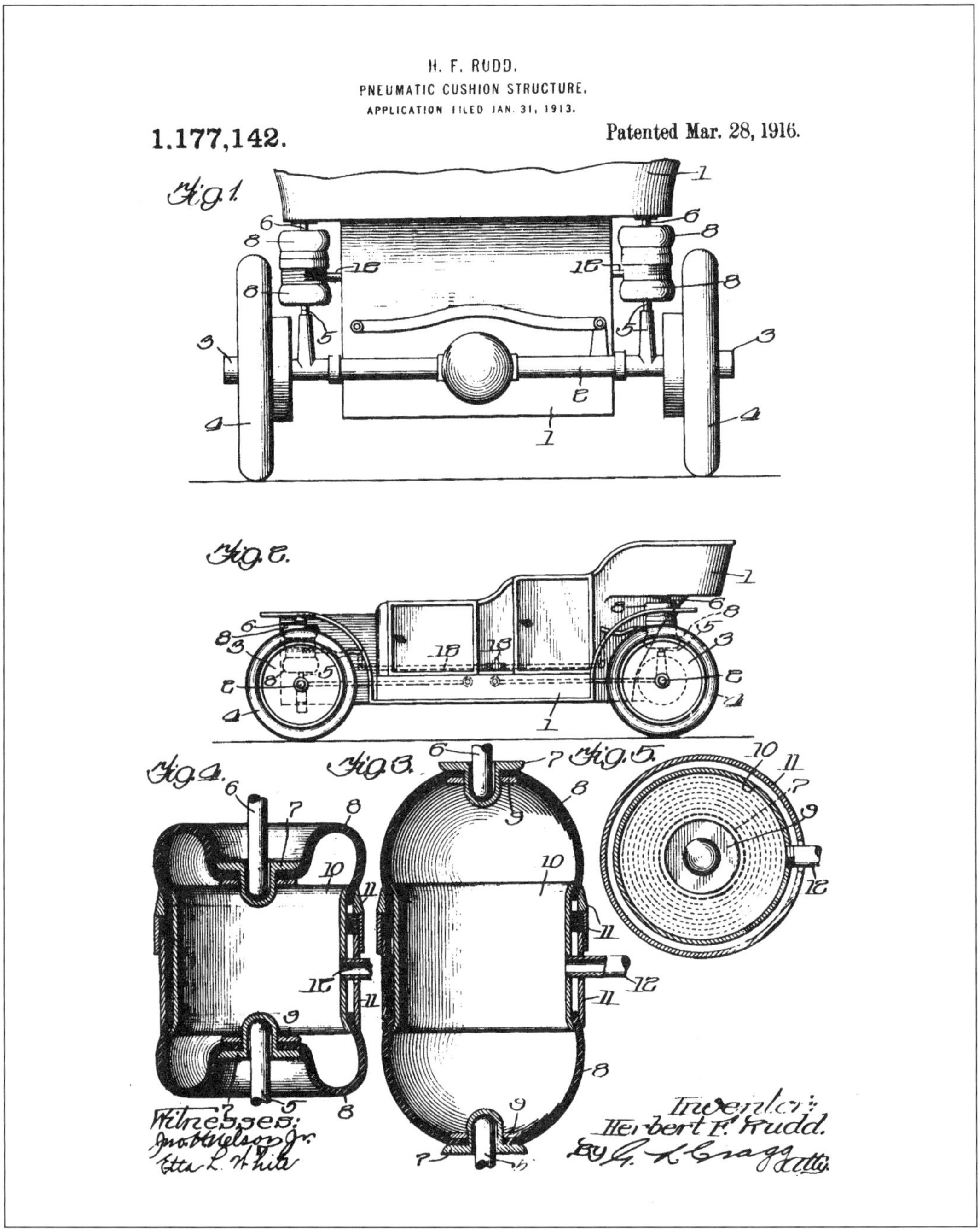

Fig. 5.20 This early air spring configuration, similar to the "2-X" or Type **C** *of the 1950s, appeared in a 1916 patent by Herbert F. Rudd.*

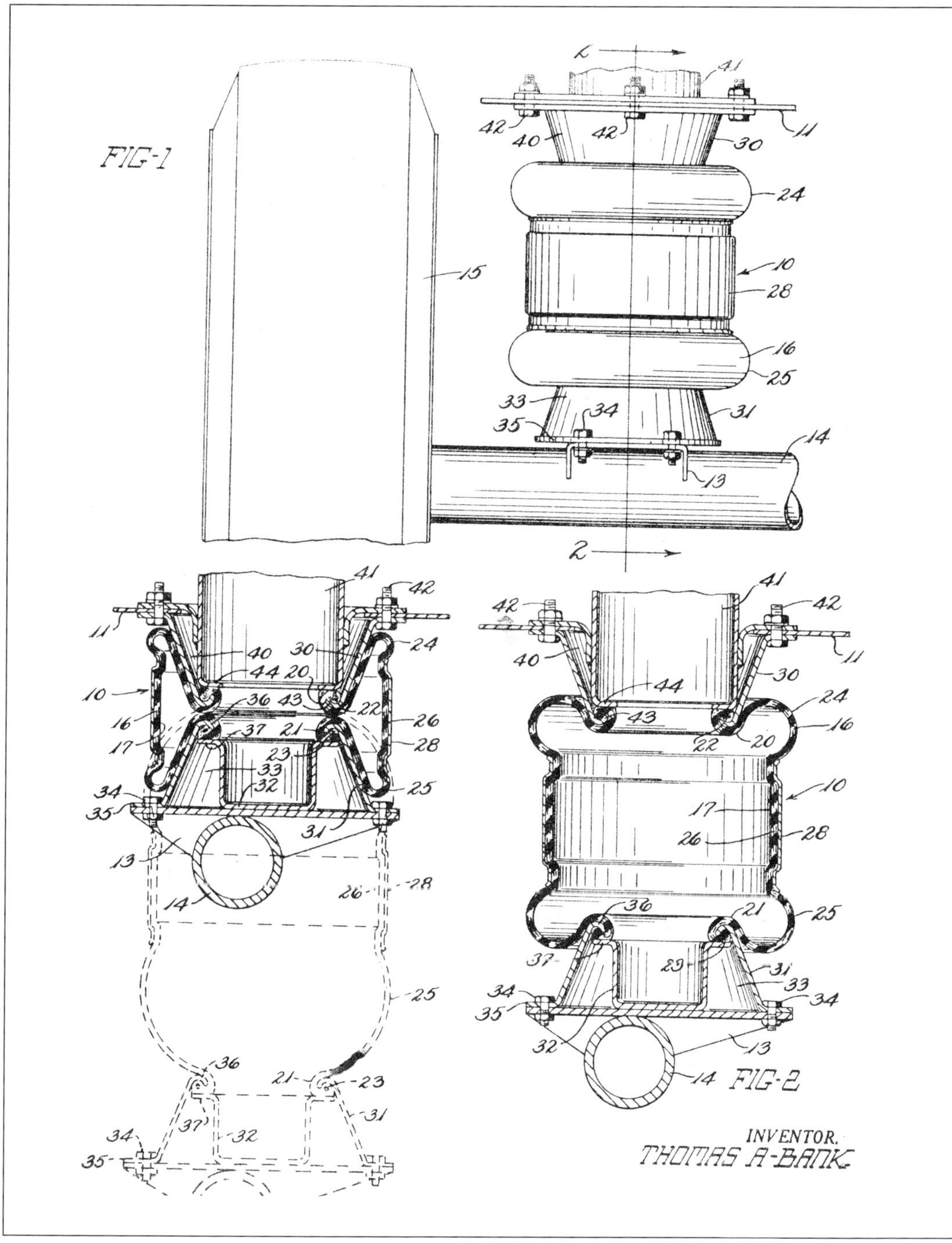

Fig. 5.21 The first Firestone 2-X, or Type **C** *design was double-ended, with equal rolling diaphragms and pistons top and bottom, as seen in this 1956 (unfiled) Bank patent application.*

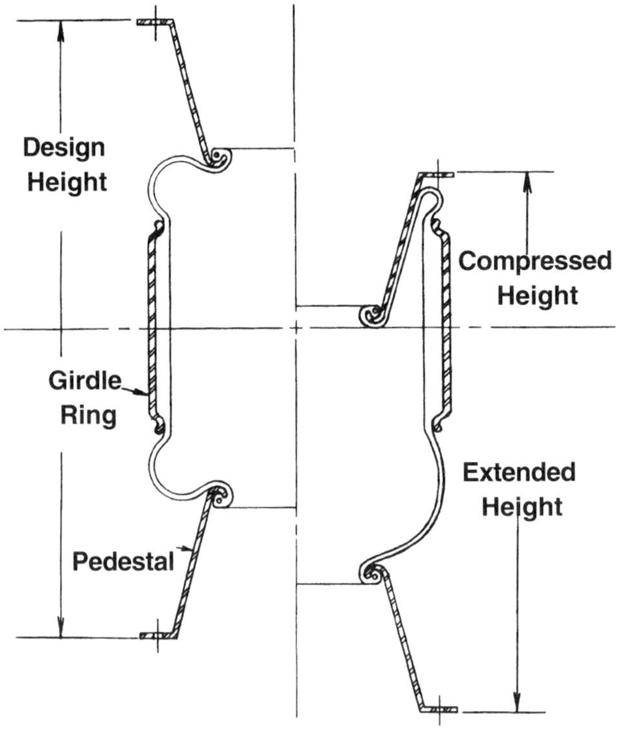

2X Type Airspring Bellows

*Fig. 5.22 The early (doubled-ended) 2-X, or Type **C** design as disclosed in a 1957 SAE paper by Firestone engineer Herbert H. Deist.*

2X Type Air Spring

Fig. 5.23 Early vertically symmetrical 2-X air spring as molded—without the wide girdle band that was assembled after cure.

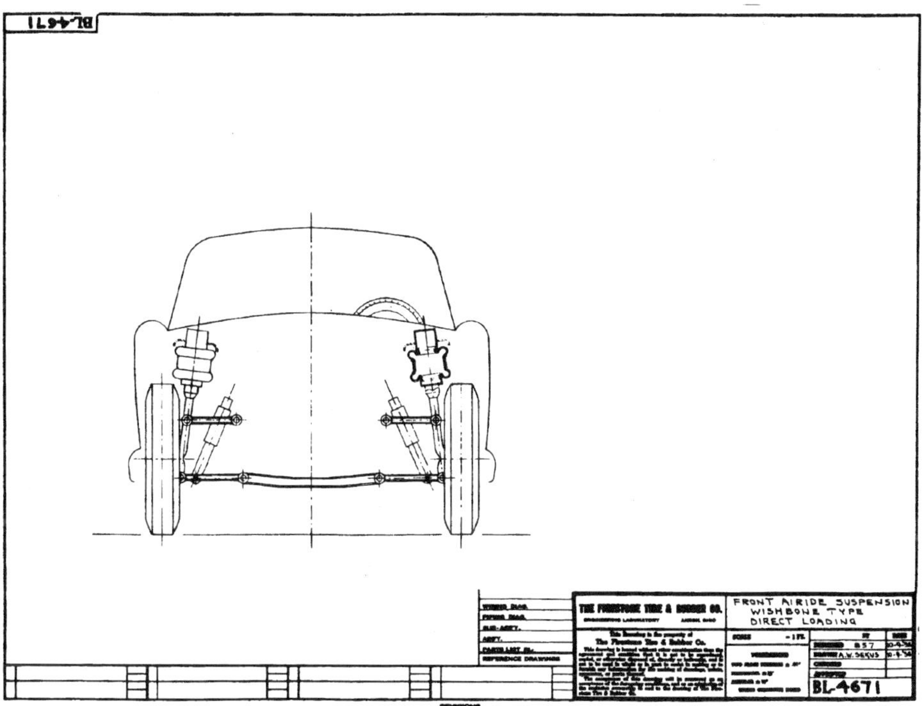

Fig. 5.24 A 1956 Firestone proposal drawing showing the 2-X design on a Ford front suspension.

Figure 5.24 is a 1956 Firestone proposal to Ford, laying out the front end of the car with original 2-X design (early C-type) air springs in place. However, Firestone's commitments to General Motors at the time (which some at Ford may have regarded as a conflict of interest), and Firestone's total absorption with gearing up to make passenger car springs on a high-volume basis (while simultaneously trying not to lose its bus business, a failed objective!) provided an opportunity for the company's competition—both General Tire and Goodyear.

Charles O. Slemmons, who had pioneered the elongated "air pillow" spring at Goodrich, later bringing it to General Tire Co., began working with Ford. And Arthur B. Hirtreiter, a new acquisition for Goodyear (see Chapter 7), began working with both Ford and Lincoln-Mercury. Neither competitor had the distractions Firestone was burdened with at the time, and the intense competition between Ford and General Motors was more than enough to overcome the traditional close working relationship Firestone had enjoyed with Ford since the 1920s.

Slemmons led an intense development effort at General Tire to help Ford perfect its pneumatic suspension system. He picked up the ball on Firestone's languishing 2-X design and ran with it. The product of this joint Ford-General endeavor, analogous to the

*Fig. 5.25 The 1958 Ford Type **C** air spring after redesign by Ford and General to eliminate the top convolution—a move that had been precipitated by frequent slippage of the girdle band.*

Firestone-Chevrolet development team at General Motors, was presented by C. F. O'Shea of Ford's Chassis Engineering Department in the SAE technical paper, "The Ford Approach to Air Suspension," delivered in January, 1958.[17]

The spring design had changed. As a field engineer in Firestone's Detroit office working on General Motors projects at the time, I remember having heard that the external girdle band on the 2-X part was troublesome—tending to slip to either the top or the bottom of the assembly. When I attended O'Shea's SAE paper presentation, I was surprised to see that Ford and General had apparently accepted the fact, and had simply eliminated the top piston.

But the reasons for the change were more sophisticated. Slemmons and General Tire Project Engineer John Welch had figured out that a lower spring rate for a given volume could be produced by providing a rolling diaphragm at one end only, utilizing the remaining space as displacement volume for the single air spring—and producing an effective area (as well as load-deflection) characteristic very similar to that of the 1-X or **B**-type spring. The resulting Ford air spring assembly, which was also used on the Mercury and Edsel, is pictured in Fig. 5.25.[16] Its articulation is detailed in Fig. 5.26.[10]

The components of the Ford spring, with the molded shape of the rubberized part now having assumed the shape of an artillery shell (substantially different from the Firestone shape above) are shown in Fig. 5.27.[16] From bottom to top they are the hourglass piston (called a "pedestal" in the O'Shea technical paper), a lock ring, the molded, reinforced-elastomer air spring, an upper end plate (bead-gripping sealing device), a mounting plate, and two "snap rings" that appear to be lock washers. The assembled air spring, (mounted on a laboratory test machine undergoing a load-deflection test) appears in Fig. 5.28.[18]

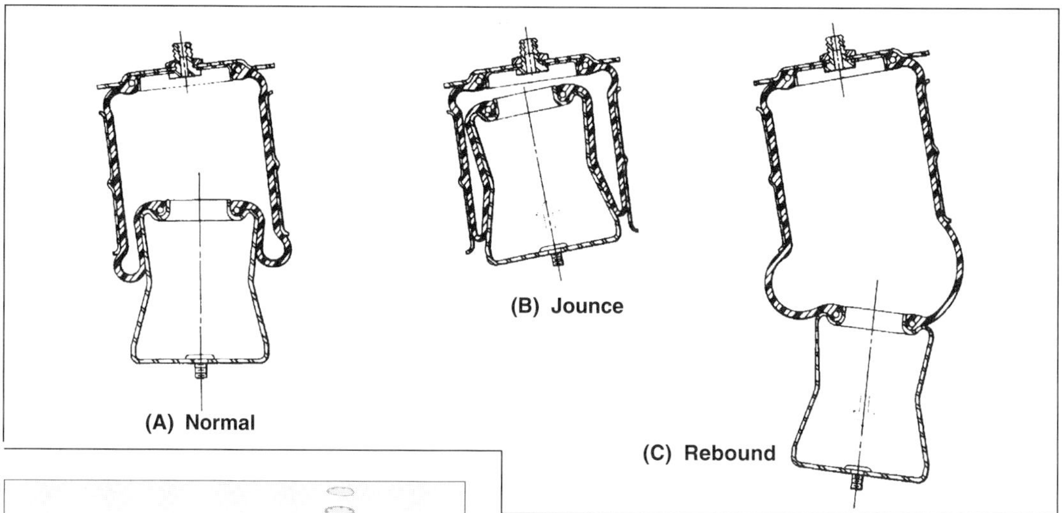

*Fig. 5.26 Showing the articulation of the Ford Type **C** air spring.*

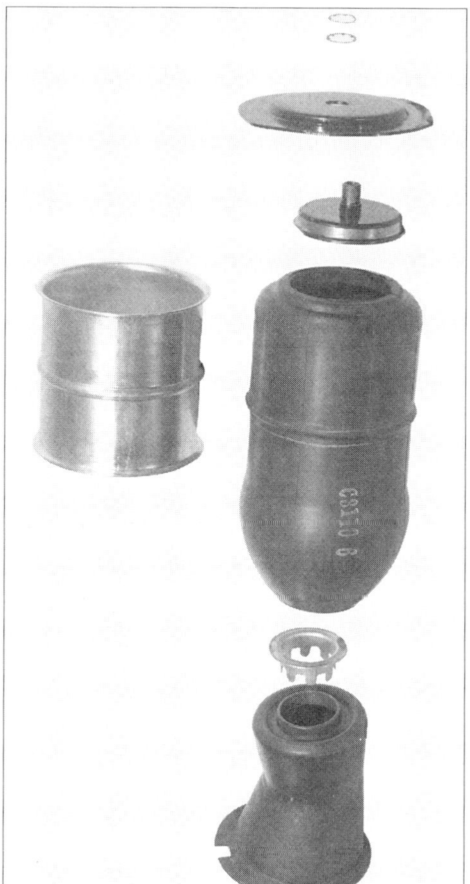

Fig. 5.27 Components of the Ford-Edsel-Mercury air spring as manufactured by General Tire & Rubber Co.

*Fig. 5.28 A Ford Type **C** air spring on test in the Goodyear laboratory.*

Other elements of the Ford pneumatic suspension system, as listed in O'Shea's paper, and shown graphically in Fig. 5.29,[17] were similar to those revealed in the General Motors SAE technical papers, but with additions:

- Air Supply System

  Air compressor and check valve

  Air reservoir [tank]

  Pressure regulator

  Filters

  Connecting lines

- Control System

  Leveling valves

  Solenoid and door actuating switches

  Restrictor valve

  T-check valves

- Air Springs

- Suspension Structural Components

  Front

    Conventional S.L.A. suspension

  Rear

    Trailing arms and attaching parts to frame and axle

    Conventional rear axle and wheel assembly

    Track bar

The single-cylinder Ford compressor, belt driven at approximately engine speed and lubricated with engine oil, had a balance pressure of 300 psi [21 bar]. Its maximum power requirement was one horsepower. The compressor fed a 400 cubic inch [6.6 liter] high-pressure air tank mounted under the right front fender. A pressure regulator valve reduced the pressure to 150–160 psi [10–11 bar] before delivering air to the leveling valves.[17]

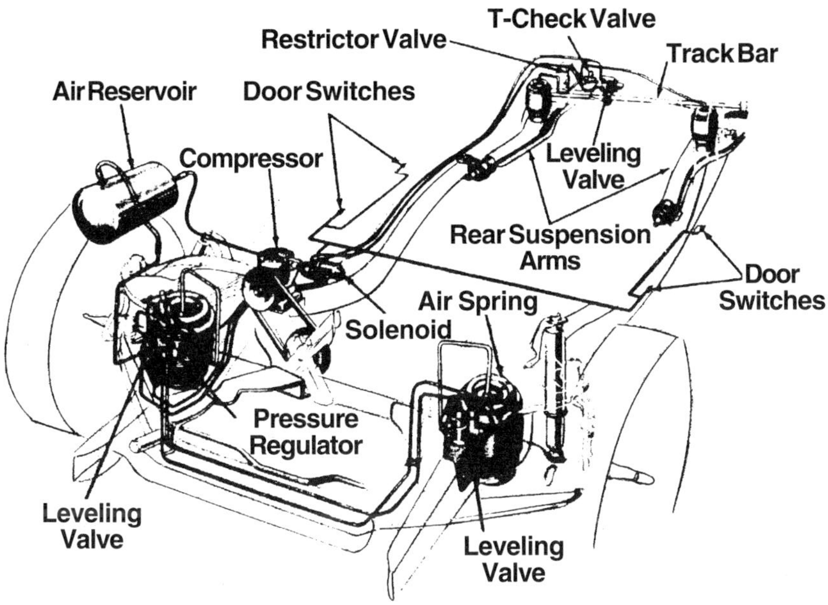

Fig. 5.29 Components of the Ford air suspension system as presented in the Ford SAE paper by C. F. O'Shea.

O'Shea explains that the reason for the higher pressure in the air tank is to "provide an air supply sufficient to operate the system when the compressor is not running"—which would not only be capable of releveling the car when passengers got in before the engine was starting, but would also compensate for any leaks in the system that could otherwise cause the car to tilt or sag while parked.

Ford engineers paid a great deal of attention to accommodation for changes in passenger load. It had a two-speed control system that "provides fast leveling when the doors are opened [tripping the solenoid dome-light switch] and load changes are likely to occur, and slow leveling when the car is in motion with the doors closed."[17]

Three height-control valves were used, one in the rear and two in front. Each had a small orifice that restricted the passage of air except when the doors were opened, tripping a microswitch that actuated a piston inside the height control valve—opening a larger channel. This secondary piston also acted "as a high pressure relief valve for the system, and will open at 400 psi [27.5 bar] if the system pressure should become abnormally high." Made of die-cast aluminum and containing rubber seals, "pistons, valves, and other components," the Ford height-control valve assembly was much more complex, and much more expensive, than the simple, rotary, two-rubber-pad valve used by Chevrolet—which had a purchase price goal of less than a dollar.[19]

There were other valves in the Ford system. "In order to provide adequate rear suspension roll resistance, the two rear springs are isolated with regard to air transfer by a restrictor valve. This valve functions as follows:

"Normally, with a single rear leveling valve, the two rear air springs are interconnected by a common air supply line. During roll, if the restrictor valve were not installed, unrestricted air transfer would take place between the rear air springs, and they would contribute no roll resistance. However, by incorporating the restrictor in the supply line to the right rear spring, air flow between two springs is severely inhibited so that the individual springs act to increase roll resistance.

"The restrictor valve is also triggered by high pressure air from the solenoid so that, when fast leveling action is required, the high restriction orifice is temporarily by-passed."[17]

In addition to the above there were "T-Check" valves: "The T-check valves between the leveling valves and the air springs are low pressure limiting devices which prevent the springs from exhausting below a set minimum value (55–65 psi [3.8–4.5 bar] front and 25–35 psi [1.7–2.4 bar] rear). They prevent extreme leveling action which would otherwise occur when one wheel is jacked up for tire changing, and they prevent complete loss of air from the springs if a leak develops elsewhere in the system."[17]

In discussing its air springs, "the heart of any suspension system," the Ford technical paper pictures an assembly (Fig. 5.30) with an "hourglass" piston, discussing how this shape affects the dynamic effective area characteristics, and how tailoring this parameter results in the familiar load-deflection curve having a low-rate center portion shown in the figure that was "exceedingly soft in normal ride movements ... followed by steep slopes toward both ends of the curve not only are necessary for acceptable handling [with such a soft, low-rate suspension] but provide excellent rough-road cushioning in jounce and rebound."[17]

The pattern of lines at 45 degrees in the above figure represents the cord angle in the finished air spring. O'Shea's detailed, knowledgeable discourse on the effect of this angle of the reinforcing nylon tire cord on air spring performance, as quoted in the following paragraph, may reflect an educational experience on the part of General Tire as the company learned how to make this new, unfamiliar product. But to be fair, whatever problems General may have encountered at the time, it must be admitted that the company later made a significant penetration into the truck market with this same Type **B** design in a larger diameter.

"The manner in which the desired rate characteristics was achieved is worthy of note," O'Shea observes. "First, it was necessary to establish the working volume for which space was available, and the effective piston area which, at reasonable working pressures, would support the vehicle over the loading range. With these fundamentals determined, the remaining factors governing performance had to be manipulated to obtain the desired results.

"Chief among these were the cord angles in the fabric layer of the rubber diaphragm and the shape of the piston. If a straight piston were employed with horizontal [i.e., circumferential] cords in the diaphragm, an almost linear load deflection curve, very

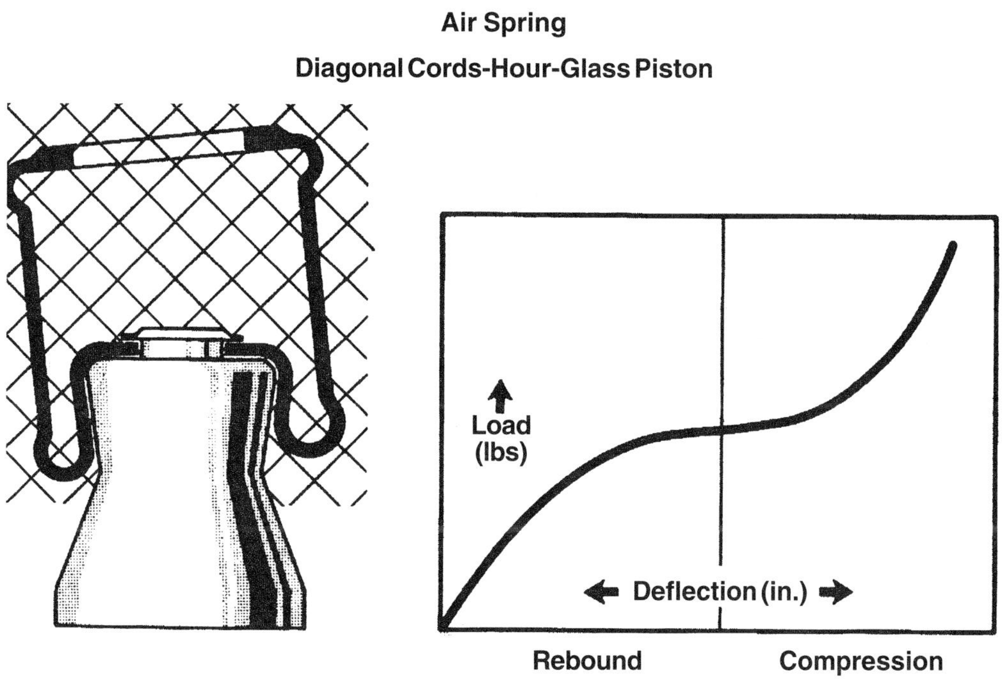

Fig. 5.30 O'Shea included this figure to suggest how critical the cord bias angle of an air spring was in tailoring both its load-deflection curve and its durability.

similar to that of a leaf spring, would result. As the cord angles are increased toward the vertical, the curve acquires an 'S' shape. If the vertical cords were to be used with a straight piston, an almost ideal curve would result. However, this is prohibited by practical limitations, such as strength and folding characteristics of the diaphragm. Contouring of the piston and using diagonal cord angles was adopted to approach the ideal curve, and to obtain ample diaphragm strength in all directions."[17]

What O'Shea doesn't mention is that the angle at which the air spring prep is built changes substantially as the part is inflated into the mold and cured. And, indeed, the cord angle also changes substantially as the spring goes through its stroke, working the bonded rubber layer between the plies in shear—particularly on **B**-type springs as the diameter of the meniscus loop (and, consequently, the effective area) changes dramatically.

Two decades earlier, Roy Brown had pontifically established 18° as the "ideal" building angle for his two-convolution bellows. (This cord angle is measured in either direction from the axis of the building drum; thus Brown's total included angle of the two intersecting cord layers was 36°.) As the prep is expanded into the mold, the angle of the cords increases substantially. The uncured rubber in which they are embedded yields in shear as the cords get farther apart—ending up with a cord angle as high as 60° at the larger diameter of a 1-X spring, and only a third of that value at the small end.

In Noblesville, when he was first given responsibility for air spring development, Tom Bank experimented with building angles from as low as 4° to higher than 30° when making 1-X *Airide* springs. Confirming O'Shea's observation above, Bank was delighted with the "ideal S-shaped load-deflection curve" that he achieved with his 4° construction. "But," as he reported during one of our telephone conferences, "when I put it on durability test in the morning before going to lunch, it had failed by the time I got back to the lab."

There was no universal "ideal" cord angle, Bank concluded. Empirical testing on both the load-deflection and durability machines dictated the optimum cord angle for each new air spring design, depending on its size and shape.[19]

Table III summarizes the relevant characteristics of the air springs offered by both the Ford and Mercury Divisions of Ford Motor Company:

TABLE III

C-TYPE PNEUMATIC SPRINGS

	FORD		MERCURY	
Property	Front	Rear	Front	Rear
Natural Frequency	50	55	49	63
Suspension Travel				
Jounce	3.7 in. (9.4 cm)	4 in. (10.2 cm)	3.7 in. (9.4 cm)	4 in. (10.2 cm)
Rebound	5.6 in. (14.2 cm)	5.5 in. (14 cm)	5.6 in. (14.2 cm)	4.8 in. (12.2 cm)
Design Pressure	105 psi (7.2 bar)	65 psi (4.5 bar)	110 psi (7.6 bar)	60 psi (4.1 bar)
Air Spring Volume	146 in.^{3} (0.9 L)	170 in.^{3} (1.1 L)	146 in.^{3} (0.9 L)	130 in.^{3} (0.8 L)

Arthur B. Hirtreiter (whose engineering career is detailed in Chapter 7) was deeply into the effects of cord angle when he developed the Type **D** air spring configuration illustrated in Fig. 5.1. In his new position as Goodyear's Chief Engineer, Suspensions, Hirtreiter had made a technological breakthrough in air spring design and construction which made it possible to achieve a low natural frequency without the expense of the metal canister associated with the Type **A** and Type **B** air springs or the bulge-mitigating external girdle band of the Type **C** unit. He called his product a "rolling lobe" air spring (for which U.S. Patent 3,043,582 was issued).[19]

Air Spring and Coil Spring Arrangement

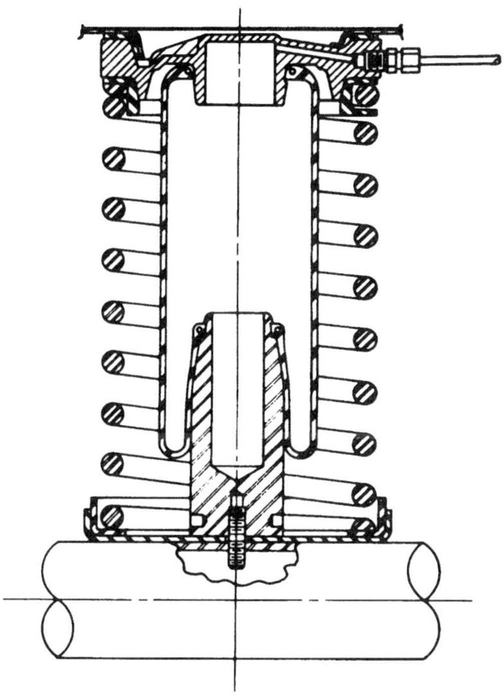

*Fig. 5.31 Type **D** air spring, conceived by suspension engineer Art Hirtreiter of Goodyear, as it appeared on American Motors' Rambler—for which application it was a long, sausage-shaped spring surrounded by (but not in contact with) a coil spring.*

The **D**-type rolling lobe, or reversible sleeve, was first used commercially on American Motors' 1958 Rambler (station wagon) suspension. In that application it was a long, sausage-shaped air spring whose length was about two and a half times its diameter (with an internally reversed meniscus attached to a piston at the bottom) mounted inside a coil spring (Fig. 5.31).[21] These air-coil assemblies were on the rear of the car only, here, as a station wagon, a substantial change in load might be expected to occur. The long, skinny air spring was not confined by the coil spring.

Hirtreiter's rolling lobe technology produced an almost cylindrical configuration without external restraint. With the vehicle fully laden and the air spring pressurized to 100 psi [6.9 bar], there remained a substantial clearance between the outside diameter of the rolling lobe and the inside of the coil spring. Fig. 5.32 pictures a pair of these American motors rolling lobes undergoing a durability test on a fatigue-testing machine in Goodyear's laboratory.[18]

In his SAE technical paper, "The Air Coil Spring: A New Factor in Rambler Suspension," Wallace S. Berry, American Motors' Director of Research, credits Director of Engineering Ralph Isbrandt for launching the company's air suspension development program, noting that, long before he joined the company's Nash-Kelvinator Division (which, when merged with Hudson, became American Motors), Isbrandt had "demonstrated an air ride car to the writer in 1938. The advantage of riding on air was obvious even then, particularly for cushioning severe bumps, but independent front suspension had just come into general use, and seemed to fulfill the need for improvement in the ride of cars of that era."[21]

(Ironically, Isbrandt had been a member of Roy Brown's staff in 1938. Predictably, the talented Isbrandt did not get along well with the autocratic Brown, and Isbrandt eventually left Firestone, ultimately becoming vice-president of American Motors.)

Berry's paper explains that a single leveling valve was "mounted near the center of the body above the rear axle." The author goes on to detail the essence of *air-coil* suspensions,

Fig. 5.32 American Motors air springs on a durability test machine in the Goodyear laboratory.

many of which would be developed for other vehicles in the years ahead (particularly on rail applications): "Most of the weight of the empty car is carried by the steel springs. … When the car is empty, 20 psi [1.4 bar] pressure exists in the air spring and each [air] spring supports about 100 lb [45 kg]. The steel spring supports 500 lb [227 kg]. As load is added, a leveling valve senses the change and increases the air pressure so that the air spring carried the additional weight of the passengers and baggage. In a loaded station wagon, the air pressure in the springs may be over 100 psi [6.9 bar]."[21]

A set of three load-deflection curves is published in the Rambler paper (Fig. 5.33) at operating pressures representing several different loading conditions. Although these are not as dramatically S-shaped as the **A**, **B**, and **C**-type springs presented above, Berry observes that, "A rising rate curve is used to minimize any tendency for [the] rear axle to bump through on rough roads. The increase in rate improves the handling and stability of the vehicle when heavily loaded, and is particularly effective for station wagons."

The Rambler suspension was a very simple system. Its pneumatics are illustrated schematically in Fig. 5.34. Berry seems proud of his product as he declares, "These advantages are obtained with only sixteen (16) air connections and a minimum of working parts." Adding, in what seems to be a dig aimed at Ford (and Cadillac), "There are no electrical solenoids and wiring to complicate the system. Reliability and ease of service

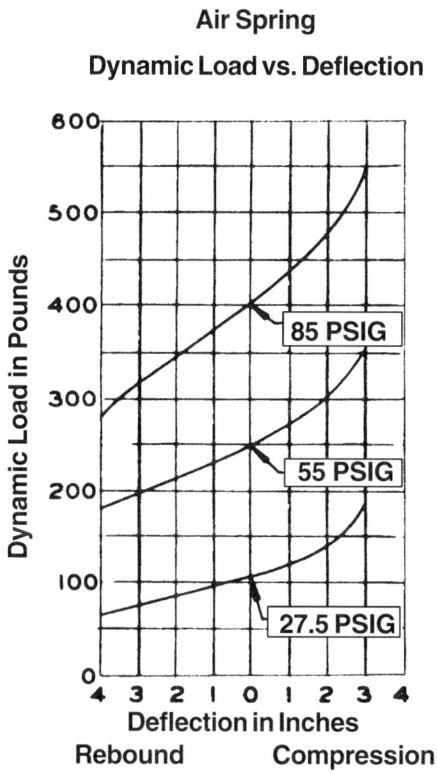

Fig. 5.33 Showing how the load-deflection curve of the Rambler air spring changes with air pressure.

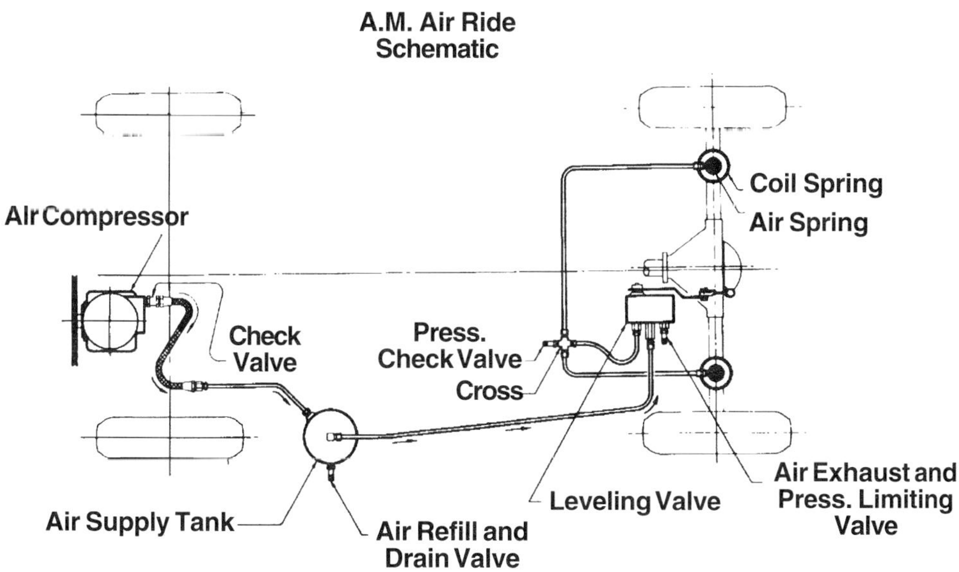

Fig. 5.34 Components of the Rambler air suspension as described in the American Motors SAE paper by Wallace S. Berry.

have been prime consideration[s] in combining the principal benefits of air springs and the existing features of the unique configuration of coil spring suspensions used on American Motors cars."[21]

Indeed, the relatively simple Rambler suspension probably was a more reliable system than most of the others whose much more complex details were first revealed at that watershed SAE Annual Meeting in January, 1958.

American Motors Rambler was not the only passenger car manufacturer to whom Art Hirtreiter had introduced his rolling lobe air spring. Indeed, Goodyear records indicate that an experimental Studebaker Champion was built by the South Bend company with a system very similar to that of the Rambler. But with Studebaker's financial problems at the time (having recently purchased Packard and, earlier, Pierce Arrow) it was not an opportune time. Studebaker demurred. The company would be out of business by 1966.

But Hirtreiter had also sewn seeds on more fertile ground. Ford's Lincoln Mercury Division built several full air suspension test cars with his rolling lobe air springs, with very promising results. In his privately-published book, *Bedtime Stories for Young Engineers*, retired executive engineer G. E. Adams was optimistic about the Lincoln system:

"We were working exclusively with Goodyear on our air-spring design, which we believed was unique in the industry. We had an outstanding senior engineer, Lou Jelsch, on this critical component and the Goodyear project engineer and their other support people knew their stuff. ... Now, an air suspension system [also] needs a pressurized air supply, a control subsystem, and leak-free plumbing. We designed plastic tubing fittings which were leak free if TLC were used at assembly. No mini-air compressors were readily available, but we found a belt-driven one that wasn't too gross, and we designed several height-control valves to complete the system scheme.

"When we thought through the control requirements to satisfy all the usual and unusual vehicle-operating conditions, it was a daunting challenge: If the height-control valve operated continuously, did it waste too much air, require a larger, noisier compressor or reservoir tanks? How do you tell it to shut off completely when parked? How fast must the system respond to changes in passenger or luggage loads? How close a height control is needed? And so on.

"It was the type of design problem which would cause you to spring awake at night, having dreamt of another operating condition which hadn't been factored into the system design. We even had a height-control-valve design which was actuated by door and trunk-lid switches, which probed the suspension heights to correct the height, shut everything off tight, then retracted out of the way! Our development engineers under

*Fig. 5.35 Type **D** Goodyear rolling lobe air spring on a 1957 Lincoln Continental test car.*

Connie Reuter made valuable contributions to the whole program.

"Ultimately, we had a car and system that worked. The Goodyear air spring met all requirements and though we weren't extremely proud of the system as a whole, it functioned satisfactorily. The best riding car I have ever been in was a 1959 Continental prototype with our air suspension."[22]

The rolling lobe spring on a Lincoln Continental test car suspension is pictured in Fig. 5.35.[18]

As will become obvious in the following chapter, Adams cautionary advice might have been helpful to other Ford as well as General Motors engineers at the time. Ultimately, Chief Engineer Harold Johnson made the decision that, because the 1958 Lincoln would be an all-new model, it would be wise not to complicate their lives by adding air suspension to the list of other innovations to be introduced on the new car. Earlier, in his own book, Adams obviously had agreed with Johnson's decision:

"The 1958 Lincoln/Continental was a formula for disaster—an all-new car. Only the wheel nuts were carryover. Plus there was an all-new plant with all-new personnel."[23]

In a 1998 letter to this author, Hirtreiter writes about how he and Goodyear also "had a long and intense but not a very fruitful air suspension program with Chrysler [with whom Goodyear had enjoyed the same kind of relationship that Firestone had with Ford, going back to the 1920s, when Goodyear was the only tire company that would extend credit to Walter P. Chrysler], and all I can say is that it was very educational for both of us.

"In the 1950s, Chrysler was using torsion bar suspensions, and the Chrysler designs were not compatible with any of the air suspension geometries at the time, so the program was really academic—though no one would admit it at the time, and Goodyear management said 'keep Chrysler happy.' So I spent a lot of time in Highland Park."

Chrysler's Director of Advanced Engineering, Otto Winkelman had some innovative but radical ideas. Winkelman had formerly been chief engineer of Mercedes Benz's Racing Division, where, in the 1930s, with other Mercedes luminaries (including Dr. Ing. Ferdinand Porsche), he designed the "Great White," a 1500-lb vehicle with a 646 brake horsepower engine!

One of Winkelman's criteria for air suspensions, Hirtreiter explains, "was to have the front and rear suspensions interconnected ala Bill Allison's Packard torsion bar suspension, or

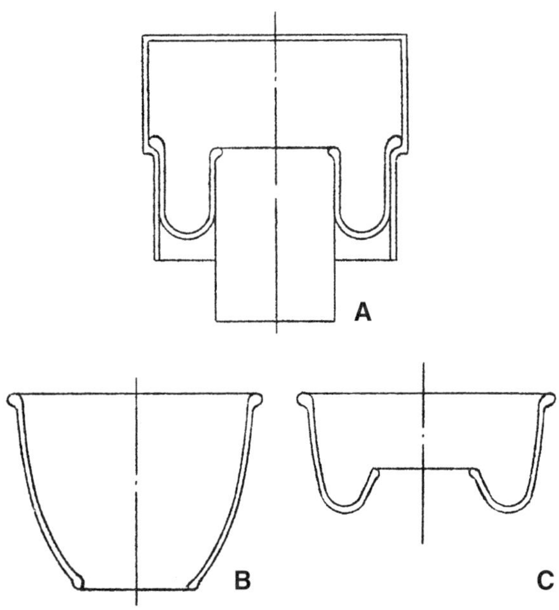

Fig. 5.36 Rolling lobe air spring configurations that Goodyear's Art Hirtreiter made for Chrysler experimental vehicles during the tumultuous passenger car years of the late 1950s. None went into production.

Molson's Hydrolastic suspension for the Mini Cooper, Austin Mini, and several other English cars. The front-to-rear connecting medium was to be air, with a complicated control valve, together with height controls. The main weakness of the system was that [compared to the European hydraulic systems] air moves too slowly, and the signal from the front to the rear of the car was too late. I tried relay valves as are used on the rear brakes of 18-wheelers. But the system was not successful—or practical—and the project was dropped."[24]

One of the early air spring configurations that was fabricated by Goodyear for Chrysler test cars (together with two of the molded shapes at the time) is shown in a 1956 sketch in Fig. 5.36.[25] The pneumatic spring design is a combination of elements of more than one of the passenger car air spring types depicted in Fig. 5.1. The long, confining circumferential support reflects Chrysler's prerequisite for high lateral stability in its air spring.

Curiously, although Chrysler (probably wisely) declined on releasing air suspension for its 1958 cars, it marketed its torsion bar suspension under the exotic name of "*Torsion-Aire*," perhaps to sound as state-of-the-art as its competitors, who were advertising the improved ride provided by their pneumatic suspension systems. Those of us in the air spring business vainly explained to outsiders that "Chrysler's air is in the *tires*!" Perhaps even more surprising, Chrysler released a simple (air-shock) pneumatic leveling system on the rear of some models the following year. Its life was brief.

At the urging of his management, Hirtreiter also spent a great deal of time at General Motors Technical Center in Warren, Michigan, and at the GM experimental facility in Plymouth. But, ultimately, after the SAE papers of 1958 came out, in a policy meeting with Goodyear's chairman Russell DeYoung, president John Gerstenmeier, sales manager Bob Warren, air spring sales Manager Andy Palguta, and Herman Morse, Director of Goodyear's Industrial Products Division, Art recommended that Goodyear stop pursuing the passenger car market and concentrate its efforts on trucks and buses, where he believed the future lay.

"This is a very expensive program," Hirtreiter remembers saying, "and honestly, neither Goodyear and Ford, Goodyear and General Motors, Goodyear and Chrysler or American Motors have enough money to make this fly. It's too expensive to go on a car.

"'Okay,' Russ DeYoung said, 'shut it down.' Andy Palguta wouldn't talk to me for two months."

His recommendation accepted, Art's department shrank from 70 employees to six—and soon thereafter, to three. But it proved to be a wise move. The small department managed to found a new, profitable business for Goodyear, permitting the company to become Firestone's major competitor in air springs, and to help expand the market for air suspension in the years to come.

Chapter Six

WILD RIDE

"AIR SPRINGS are the big technical news from Detroit for '58," reported Hubert Luckett in a well-researched feature article in the January, 1958, issue of *Popular Science*. '58 is the guinea-pig year for air suspension. If no serious weaknesses show up, look for redesign of cars."[1]

Nineteen months later, *Automotive News* would conclude that "Automotive engineers have received their most severe rebuff in many years from the U.S. consumer, as air suspension becomes almost extinct this fall. Of the 13 auto makers which introduced it in the past two years, only two will offer it on their '60 lines."[2]

Encouraged by the spectacular success of air suspension on buses, and concerned about expiration of the 1930s Brown patents (as well as what their competition was up to) both Ford and General Motors, followed by American Motors, had moved their planned release dates for passenger car air from 1959 to the 1958 model year. At the same time, because they were designing vehicles that would sell for a tenth the price of a bus (and could be expected to last only a tenth as long in terms of mileage), Detroit engineers had sought to take the high cost out of such pneumatic components as compressors, height-control valves, and even air lines. It was a recipe for disaster.

In less than three years pneumatic suspension on passenger cars had gone the way of Whites and Locomobiles (steam cars of the early 1900s), and it would take almost thirty years for full-air suspension to successfully return on passenger automobiles. *Automotive News* went on to recite the hard facts:

"Air suspension never did sell as well as expected. Sales on the high-priced, low-volume Brougham revealed nothing. The best year was 1958, when total industry sales were about 100,000 units. Cadillac had it on 14 percent of its cars that year, followed by Buick, 10 percent; Oldsmobile 6 percent; Pontiac 3 percent; Chevrolet 2 percent; and Ford and Rambler, both less than 1 percent. ... Despite some improvements, the air spring really fell on its face in 1959, declining to less than 50,000 [car] units.

"This rebuff by the U.S. consumer, which in many ways was the worst ever experienced by automotive engineers, was brought about by several factors. Chief among these, according to many informed people, is that merchandising people overruled the engineers, forcing the debut of the [system] before it was mechanically satisfactory and before enough cost was removed.

"Typically, one engineer said, 'If only someone had had the courage to tell management that [air suspension] wasn't ready yet.' ... Some systems had as many as 28 possible leakage points."[3]

As one who was on the firing line at the time—and sometimes it did, indeed, seem like a war—I agree with the *Automotive News* assessment, but I think there were *two* major reasons for the failure: haste and *secrecy*.

Sometimes setting up their drafting rooms in what had been warehouses or former equipment-storage space, R & D development groups working on air suspension were completely isolated from the mainstream. The groups had almost no contact with the engineers designing production cars with steel spring suspensions. Vendor representatives without a special pass were forbidden. The objective, of course, was to prevent any details—even the very existence of their programs—from leaking to competition.

Luckett put his finger on one of the resulting problems in his prescient Popular Science article:

> "The contrast between the ride of a standard '58 car and one with air springs is very slight for normal driving. [And] subtle differences that show up when the cars are hard pressed may be due as much to other suspension changes in 1958. ... Before you plunk down that extra $125 to $200 or so for air bags, make a point of trying a steel-spring version of the same '58 car. One good comparison is the Chevrolet. ... "[1]

The natural frequency of Chevrolet's rear pneumatic suspension was 64 cpm—significantly lower than that division's norm in the past. But 1958 was also the first year that Chevrolet put coil springs on the rear. Fully loaded, the natural frequency of the rear suspension of cars with *steel* springs was 60.

Motor Trend's Detroit editor, Joe Wherry, after driving both steel-spring and air-suspended Chevrolets on GM's Milford Proving Ground was obviously disappointed: "The new air suspension, good as it is, may still take a back seat to the four coil springs for some time." Wherry also complained about the handling of the Bel Air with air suspension.[4]

He obviously liked Ford's performance better: "A level ride and constant car height despite the load inside, and baby buggy smoothness on lumpy roads are achieved with the optional "Ford-Aire" suspension system."[5] Wherry's subjective evaluation is borne out by the published data on Ford's (and the Edsel's) air suspension, whose natural frequency on the rear was 55 cpm. Mercury's, surprisingly, was 63 (with a value of 49 on the front—essentially the same as Ford-Edsel's 50 (see Table III, Chapter 5).

Leading his cluster of articles in the December, 1957, issue of *Motor Trend*, Wherry gave the highest marks to Buick: "[The] air suspension on '58 Buicks, optional on all models, is one of the best offered in the industry. ... Most surprising quality of the new air-suspended Buick is its ability to better previous models in overall handling and roadability. Fast acceleration from a standing start causes virtually no tail end squat; panic stops from high speed cause less nose diving than expected, and stability when cornering fast is remarkably good [a remarkable comment recognizing that Buick had the very lowest natural frequencies offered on any car that fateful year]. ... If new styling and superb riding ease will sell cars—and we tried the rear seat too—we predict a rosier '58 for Buick."[6]

But, partly because of their considerable mass compared with modest changes in their payload, on many 1958 cars the only obvious improvement in performance that air suspension had to offer was leveling—which Luckett appropriately applauded, relating it to the radical changes in styling that had ensued after World War II:

"Air suspension is not just a sales gimmick. The fact is that air suspension (or something like it) is inevitable if the present direction of car design continues. The frets and inconveniences that have been aggravated by the trend to lower car heights and softer rides are many. Unloaded, the car stands at the curb with its rear end up in the air like an unballasted tanker. Loaded, the headlights search the treetops (or blind other drivers) and rear vision is shut off. Loaded, it is too willing to bottom. Loaded or not, on even moderately abrupt inclines, the bumper guards may drag bottom."[1]

But, as *Automotive News* pointed out, with few exceptions, these were pneumatic systems that had been developed in haste—systems that would be repented for the next three years. Each fearing that its competitor might beat its own release date by a model year, both Ford and General Motors moved their 1959 release dates to 1958. Many system components, as a result, were insufficiently developed. Testing was inadequate. At Firestone, although there were no major problems with the air springs themselves, our drafting room was almost constantly engaged in designing new molds or modifying existing ones to tweak load-deflection curves in order to improve ride performance.

And there was pressure to take *cost* out of the systems. Some of it was unrealistic. GM Truck & Coach Division had paid its sister Delco Division $33 apiece for leveling valves ($99 for the requisite three), obviously much too steep a price tag for an option that was to sell for $125 (a price that would also include four air springs, a compressor, an air tank, and associated plumbing). So purchasing objectives were established: at Chevrolet, $1.50 each for air springs; a dollar each for height-control valves; less than $100 for the compressor.

In the isolated R & D garage at Chevrolet it became a routine, and a cynical inside joke, that whenever it was known that assistant chief engineer Harry Barr was on his way, there would be a scramble to start up all the air suspension cars—which would otherwise be tilted at bizarre angles when he walked in.

"I can't imagine anyone in his right mind releasing that option at that state of development," one of the engineers intimately involved with the project at the time told me forty years later.

At the end of 1958, when Fred Haushalter retired, I inherited his company Ford (which, of course, had air suspension) along with his job as Senior Automotive Engineer. To get thoroughly familiar with the performance of the car, I took it on a Florida vacation with my family that summer. Because of the demands of my job during that first year of passenger car air suspension, we had to delay our vacation until the 4th of July—not an ideal time to drive to Florida in a car without air conditioning.

Operating at 300 psi [21 bar], the air compressor frequently got hot enough to char its lubricating oil, and it eventually coked up sufficiently to cause the compressor belt to slip—even after retightening. The resulting belt noise, some of it above the range of human hearing, became a continuous pulsating shriek. As we drove through one small town after another (in an era before the construction of interstates), all of the dogs in the community would flock to both curbs, worriedly watching us go by—perhaps suspecting that we had one of their number trapped under our hood.

As described in Chapter 5, Ford had a very sophisticated, very complex valving system, in which air line condensate frequently froze during the winter—much to my chagrin on those cold mornings when I drove my company car to *General Motors* accounts. I learned to take lots of joshing from GM engineers on cold mornings every time I rolled in with my Ford's posture cocked at odd angles.

To avoid similar freezing problems, Chevrolet engineers had devised what seemed to be a neat, simple solution. As reported in the Hansen et al. Chevrolet SAE paper, "a simple alcohol evaporator at the air intake [behind the warm radiator] provided the needed protection."[7]

Unfortunately, the bonding cement on the tiny rubber sealing pads in the Chevrolet leveling valves proved to be soluble in alcohol. If the pad sealing the exhaust orifice was the one that dropped off, the air spring(s) served by that valve would collapse, the suspension would sink down onto the jounce bumper, and the system would continuously leak air. But if the pad sealing the high-pressure inlet came off, one or more air springs would inflate into full rebound, where they would be maintained at a steady pressure of 225–250 psi [15.5–17.0 bar] until one of them eventually burst—since the Chevrolet system had no pressure-reduction valve between the high-pressure tank and the leveling valves.

Although our 1-X type (**B**-type) air springs were built to have a burst pressure in normal design position of several hundred pounds per square inch (e.g., 17 bar), and at least twice that amount in full jounce (where the highest pressures might normally be expected to occur), when they were distended for long periods in full rebound, permitting the nylon cords to stretch, their burst pressure could be as low as 175 psi [12 bar]—or even less if they were kept in tension long enough. There is sufficient energy packed into an air spring at these pressures that I can knowledgeably (as a former field artillery officer) compare the resulting report to the muzzle blast of a 105 mm howitzer—a noise more than a little disconcerting to a driver whose car had already assumed a wild, rakish posture as he tooled down the highway.

These were not, unfortunately, isolated incidents. I even encountered them on weekends. Sitting in my living room in Birmingham, north of Detroit, one Sunday afternoon in the fall of 1958, I heard the familiar explosion issue from a Chevrolet driving north on Adams Road about a hundred yards from my house on Villa.

I hurried over to console the frightened driver, who turned out to be the daughter of a Chevrolet executive in Flint who had lent her his company car for the day. With her right front suspension resting permanently on the rubber jounce bumper, I explained what had happened. When she understood that my company had made the spring that had failed, she plaintively asked if I could get her a replacement. I told her I had at least one 1X24A Chevrolet spring in my desk at the office, and I'd be glad to go downtown and get it and even help the neighborhood filling station put it in, but I didn't think she'd make it back to Flint without another burst unless she had a new leveling valve. So she limped back home, her car bowing obeisantly for fifty miles—less than a fan of air suspension.

We developed related problems at Cadillac. In switching from GM Tech Center **A**-type springs on the 1957 Eldorado Broughams to **B**-type 1-Xs on 1959s, Cadillac had retained the "Air Lift" lever on the dash, which permitted an owner to demonstrate that he had purchased air suspension by inflating the springs into full rebound. There probably were rare situations in which this option was useful, but it was essentially a gimmick—one that caused more problems than it solved.

On one occasion, Cadillac engineering personnel responded to a call from Spain in an effort to determine why a wealthy, influential Cadillac owner in that country was breaking suspension parts on his limousine at a prodigious rate—actually pulling ball joints out of their sockets. It seemed the man was a grouse hunter, and he had discovered that if he pulled the Air Lift lever, he could drive his Cadillac over open fields and plowed ground at high speed without bottoming out.

There was never any deliberate attempt on the part of Cadillac engineers to conceal the existence of the Air Lift lever from Firestone. But in the prevailing atmosphere of secrecy that surrounded all of the new features on the new cars, the matter simply never came up. If it had, we would have objected, pointing out that, unlike **A**-type springs on the Eldorado Brougham, the **B**-type 1-X design did not have the external supporting skirt around the air spring that the Tech Center spring had, and, consequently, its burst pressure was significantly less in full rebound—as we had already learned at Chevrolet.

The hazard was aggravated at Cadillac by a decision to increase the regulated air pressure delivered to the leveling valves in order to fill the springs faster when passengers got into the car. Then there was the tolerance spread on the pressure regulator valves, as well as on the bell-curve of building quality of the individual air springs.

As it turned out, it was the car jockeys (workers who jumped into the cars as they came off the line and drove them up into the Cadillac warehouse) who inadvertently instituted a test to weed out any marginal systems. These young men found that they could shorten their cycle time if they pulled the Air Lift lever before speeding up the steep warehouse ramps; and most left the Air Lift lever pulled out after they parked the car and hurried downstairs to pick up the next one.

It was the company's night watchmen who, involuntarily, became the inspectors who spotted those cars having an unfortunate combination of tolerance spread in their pressure regulators and/or of air spring quality. On the night shift, unaware of the gradually stretching nylon cords in the sea of Cadillacs through which they threaded their way in the semi-darkness of the warehouse, these unfortunate souls, some of them World War II veterans, would periodically be scared out of their wits by an explosion that sounded quite a lot like an artillery shell. Their union had a legitimate grievance.

Although as a field engineer I had been called on customer carpets more than once because of emergent factory problems over which I had no control, if I had not been there, I would find it hard to believe that this matter was finally resolved in a conference that took place at midnight on a Sunday night in the office of Cadillac's Director of Purchases. After the meeting, Firestone executives and I checked into a downtown hotel until morning.

To satisfy Cadillac management that we were not shipping defective air springs (and perhaps as penance), a staff engineer and I were required to spend a day in the laboratory running burst tests on air springs with *air*—in an era before hearing protection—because purchasing personnel wouldn't believe that bursting with water pressure could yield the same values. I learned later that so many ceiling tiles had been loosened that day that the ceiling of the lab had to be redone. The experience was not a boon to my field artillery deafness.

There were a number of related reports from the field, one of which I heard about from Firestone's Law Department: It seemed that an elderly Cadillac owner in Florida, who was washing his car in his driveway, had pulled the Air Lift lever to facilitate washing his wheels. When one of his air springs exploded, it startled him so badly that, jumping back, he tripped over his water bucket—sprawling across his Zoysia-grass lawn. His wife, hearing the explosion, rushed to the window, where she was shocked to see her prostrate husband, apparently mortally wounded. Both were taken to a hospital emergency room, where they each were given a sedative before being sent home.

The occasional humor kept us sane. Once, while developing the Pontiac system, suspension engineer Bob Lovelace, his test driver, and I went to lunch in a prototype car on whose front suspension we had managed to depress the natural frequency to 40 cpm. The car rode like a yacht as we drove through the streets of Pontiac on the way to the restaurant.

What we didn't know at the time was that its nearly-flat dynamic (adiabatic) load-deflection curve would become *negative* isothermally—until we came to red traffic lights, that is. Each time we stopped, slowly sliding backward down the negative load-deflection curve, the front end of the car would humbly bow down a total of 4.1 inches [10.5 cm], surprising drivers in adjacent lanes. Then, as the height-control valves kicked in, the front would suddenly rise up a total of 8.9 inches [22.6 cm]—to the astonishment of the drivers of nearby cars whose attention we had already attracted—thereafter sinking all the way back down to repeat the cycle. Since the very existence of our air suspension development was confidential at the time, we could offer our neighbors no explanation. We just kept a straight face, pretending that was the way our car was supposed to work.

By the end of 1958, a modest amount of tinkering with the air suspensions had taken place. "Suspension systems continue to draw a major share of engineers' attention," the December, 1958, *SAE Journal* headlined. "This year Buick abandoned air springs in front; Chrysler offers optional air springs in the rear for leveling. Cadillac's freon-filled shock absorbers are the first to be used in U.S. cars."[8]

Because, as originally released, the ride of Chevrolets with air suspension was virtually indistinguishable from cars with coil springs, the *Journal* noted that, "There have been some modifications in Chevrolet air suspensions for 1959, including reduction of front spring rates of 40% and a decrease in rear air spring rates of 15%. Other changes include a change in front and rear piston contours [going to an hourglass shape to reduce the rate], new front bellows, and minor changes in valving [which, as reported above, seem to have been indicated—especially if alcohol antifreeze would continue to be introduced into the system in winter!]. The net result of these design changes is a much softer, smoother ride."

The same article reported that "The 1959 Buick equipped with air suspension has two coil springs in front and two air springs in the rear with individual levelizer controls. ... Designed spring rates at the rear are approximately 55% of coil spring suspension for a distance of 1 in. [2.5 cm] on each side of normal design height. ... Variable spring rate at the rear becomes effective if the wheel moves into compression or rebound beyond 1 in. [25 cm] travel. Compression travel has been increased to 5 in. [12.5 cm] in the rear. ... In addition to a soft, boulevard ride, the Buick suspension, it is claimed, provides superior roadability on winding roads."[8]

The changes suggest that Buick may have made the changes to satisfy motion sickness complaints, which were rumored at the time. I also remember hearing at the time about Buick test cars being run with and without sway bars, as well as about experiments with "hop dampers"—quart-size metal flasks attached to the axle containing a weight sandwiched between two coil springs.

Despite the problems with air suspension encountered by its competitors, Chrysler apparently felt pressured by the same forces that had caused its torsion-bar suspension to be inappropriately named *Torsion-Aire*. The same December, 1958, *SAE Journal* article on changes in passenger car models for 1959 reported that "Chrysler's new rear suspension automatic leveling control combines the features of the Torsion-Aire suspension with the advantages obtainable from air chambers. Flat cornering, level attitude during acceleration, and freedom from nose-dive during stops, all features of the Torsion-Aire suspension, are added to automatic leveling and nearly constant ride frequency, it is claimed."

The article goes on to detail the placement of the engine-driven air compressor and its operating pressure (220 psi [15 bar]), as well as a single height-control valve which admitted air at up to 110 psi [7.5 bar] to a low-pressure tank attached to "air chambers" on the rear axle. These "air chambers" seem to have been air shocks. There was little further publicity about the system, and no production release.

Looking back at that period forty years later, retired Cadillac Staff Engineer John Hoban had several observations about "non-safety-related bugs found in a premature design that were just too numerous for Manufacturing and Service to digest and satisfy customer expectations. The air suspension disaster was further compounded," he said, "because all divisions of GM (*and* our competitors) rushed into production at the same time. Our customers soon tired of air leaks that could not be handled satisfactorily in the field. I believe that if GM had started exclusively at Cadillac with our 1959–60 design, air suspension would have won the customer respect necessary for its survival."[9]

It was a terrible three years for all of us—car manufacturers and suppliers alike. I developed a duodenal ulcer which didn't heal until 1960—by which time, in an O. Henry-like twist, my company had sacrificed the bus business it had enjoyed since 1953 in a desperate effort to preserve its newly-acquired, hopefully much-higher-volume passenger car business.

It was an odd time to be offered a promotion, and when first asked I turned it down. But Vice President Russ Mitchell was an excellent salesman (which is why, after all, he was Vice President of Sales), and he persuaded me to move my family to Firestone Industrial Products Division's headquarters in Noblesville, Indiana, as the company's first Product Manager for *Airide* springs.

It would be a career gamble: If we achieved a million dollars in sales by the end of the 1960 fiscal year (which would be the end of October, only nine months away) the company would stay in the air spring business and Tom Bank would begin converting the high-volume passenger car presses that had been developed by Herb Deist and the Engineering Laboratory in Akron to our altered new-business needs. If we failed, the Airide Plant would be shut down and employees would be laid off, probably including this Product Manager.

Chapter Seven

NEW ERA

Firestone's loss of its General Motors bus business was due in major part to Goodyear's acquisition of a talented engineer the rubber company had lured from the truck industry, where, for many years, he had developed new suspension systems. A 1936 graduate of Polytechnic Institute of Brooklyn, Arthur B. Hirtreiter, P.E., had already carved out two distinguished engineering careers, most recently five years at Reo Motors as Assistant Chief Engineer for Commercial Vehicles, and 24 earlier years at Mack Truck—where he was, in turn, Suspension Engineer, Metallurgist, and Stress Analyst, becoming Mack's Truck Engineer in 1939, serving in that position until he was lured from Allentown in 1950 to Reo's headquarters in Lansing, Michigan. At Goodyear, Hirtreiter was appointed Chief Engineer for Suspensions, Project Development, and Advanced Systems, and within three years he had become Chief Engineer of the company's newly-established Air Spring Division.

Interestingly, Hirtreiter remembers Roy Brown's calling on him at Mack Truck in Allentown, Pennsylvania, in 1936, when he was Mack's Suspension Engineer. Art's visitor was not pleased when, although he listened carefully, his host found nothing of interest in Brown's air spring proposals at that time. Neither Brown nor Hirtreiter could have imagined then that, two decades hence, Art would be sought by Firestone's arch rival to develop Roy Brown's product. As Chief Engineer of Goodyear's Air Spring Division, Hirtreiter would invent a revolutionary new air spring design. It was Hirtreiter's early expertise in stress analysis at Mack Truck that would enable him to respond to GM Truck & Coach Division's need for a radical new concept.

"We want the 'New Look' coach to ride like a Cadillac," GM's John Killinger and John Locklin had told Art. This ruled out retaining the two-convolution Brown-configuration air springs, which, even with an *infinite* displacement volume, were incapable of producing a natural frequency less than 75 cpm. Besides which, as related in Chapter 2, some of GM's welded steel air beam reservoirs were already rusting out in coastal areas or in northern cities with heavily salted streets in winter. Some transit systems had resorted to sealing off these leaky displacement volumes with a steel plate between the air spring bead and the air beam flange, elevating the natural frequency of their buses to about a hundred cycles per minute, defeating the ride-comfort and shock-isolating advantages of their pneumatic springs.

Because GM Truck & Coach had specified a natural frequency of less than 60 cpm for the new buses, Firestone had proposed its **B**-type 1-X design (see Fig. 5.1)—the air spring configuration that was, at that time, being designed into GM's first highway truck-tractor with air suspension, the DSR-8000. Such **B**-type 1-Xs would, of course, require tooling for the steel displacement canisters into which the air springs would fit.

"We had to come up with something different," Hirtreiter explains. And he did. He conceived the Goodyear "rolling lobe," the Type **D** air spring, as detailed in his U.S. Patent No. 3,043,582 (Fig. 7.1).[1] Although capable of natural frequencies in the passenger car ride range, Hirtreiter's rolling lobe requires neither the massive displacement reservoirs of Types **A** and **B**, nor the external restraining band of Type **C**. Like the first two, the rolling lobe has an inflated loop, or meniscus, which rolls up and down a piston, but above this loop the outside diameter of the air spring retains its cylindrical shape without any external restraint.

I've been a friend of Art Hirtreiter's since his Mack Truck days. When we first discussed this book last year, I asked Art how he happened to think of the rolling lobe concept, and, while he was at it, could he please explain *why* it works. I'm glad I recorded our conversation:

"As a part of my indoctrination into the rubber business when I joined Goodyear (I really knew nothing about it at all)," Art modestly explained, "my mentor was Elmer Kimmich, a brilliant engineer-mathematician who had been the youngest man ever to graduate from Case Western Reserve. I became interested in his special area of development, which, a few years earlier, had been pressurized *hoses*—especially power steering and brake hoses.

"Up to that time, the hydraulic brake hoses that were used on the Duesenberg and early Chryslers, and some of the others, had been wound with wire so they wouldn't expand when the brakes were applied. But Kimmich somehow figured out mathematically that if he *braided* a hose which had a braid angle [off the centerline of the hose axis] that was the arc tangent of the square root of two—which happened to be 54 degrees, 44 minutes (I'll never forget that) he would have an *equilibrium* angle that would prevent the hose from expanding.

"Kimmich then made some hose that way, tested it, and found it stayed the same diameter—didn't change at all. That was the start of the technology for all the hydraulic brake hose that's now used worldwide.

"Then when power steering came in, they had all kinds of gadgets to subdue the pulsations from the hydraulic pump (most of them were vane pumps and they had a pulsation). So then, to get away from using air tanks and what-not, he braided some hose at an angle of *less* than 54° 44′ so, as the pulsations took place and as the hose was

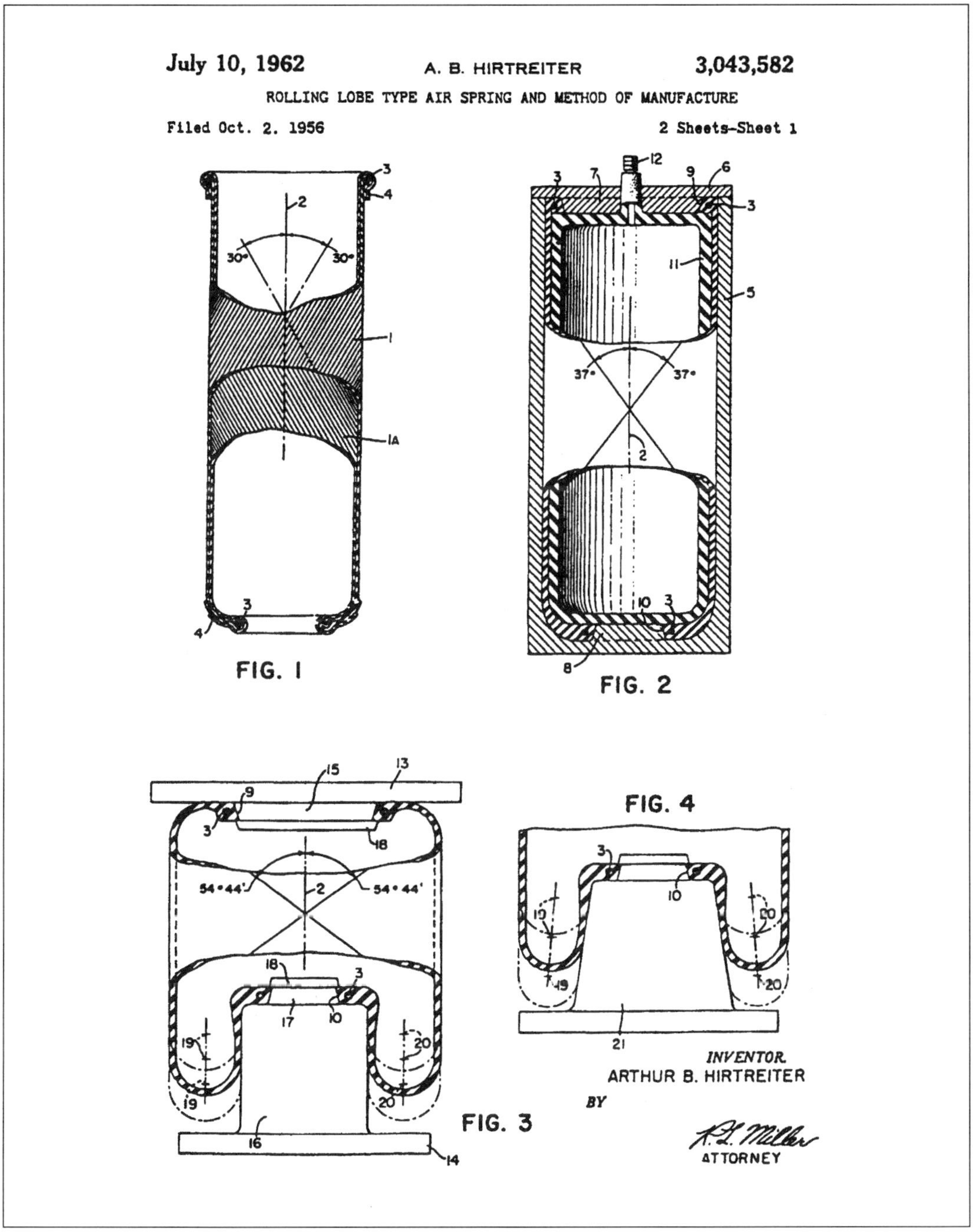

Fig. 7.1 The original Hirtreiter patent for the Type ***D*** *"rolling lobe" air spring. Although capable of natural frequencies in the passenger car ride range, the rolling lobe design requires neither a displacement reservoir nor an external restraining band.*

intermittently pressurized, it expanded slightly, coming to this same 54° 44′ angle—which made it a larger diameter, but then it didn't change any more.

"Kimmich also mentioned that, in making certain kinds of *fire* hose which is used on extension ladders, these are built at an angle of *greater* than 54° 44′ so when the water is turned on and hose is pressurized, the diameter *decreases* and the hose becomes a little *longer*—so it doesn't pull the fireman backwards off the ladder. But the equilibrium angle was the same: 54° 44′.

"He told me about this and a lot of other related things, and I thought, 'gee, I'll bet we could make an air spring that remained a cylinder at its large diameter, and still would be able to work around a piston at the other end at the small diameter.'

"So, I made some sketches, and sent them to my friend Bob Henning at the Marysville Plant. He made one up, and Elmer Kimmich said, 'I think that'll work.' But he was the only one at Goodyear. But they didn't realize I was a hard-headed, stubborn Dutchman!"

Henning recalls that the first rolling lobes he made had pistons at both ends like one illustration in Hirtreiter's patent.

"After I analyzed it," Art said, "I realized that the shear loads on the rubber materials were much higher than on belts or hose or tires or anything else, so I went over to Goodyear Research and they came up with a very good—and very expensive—Neoprene compound. And that worked out well."[2]

These shear loads, as well as the shear strain, or stretch, in rolling lobe (or "reversible sleeve," **D**-type) air springs are due to the large change in cord angle as the spring operates on the vehicle. The cords "pantograph" as the diameter of the spring changes by a factor of two from the major diameter of the spring to that of the piston and back again, hundreds of millions of times in service as it works up and down through its stroke, stretching the thin layer of rubber between the cord plies to a much greater degree than in a tire sidewall, for example—or in a two-convolution bellows-type air spring.

To achieve durability on his new configuration that would match the service life of convoluted bellows, with their lower stresses, self-described "stubborn Dutchman" Hirtreiter insisted upon, and with an assist from no less a fan than Goodyear's Board Chairman Eddie Thomas, managed to get *knotless* tire cord—an unheard-of commodity in the company's tire plants up until that time.

The second page of Hirtreiter's rolling lobe patent (Fig. 7.2) discloses a wide variety of piston shapes, some of which later became components of production Goodyear air spring assemblies. His first design concept, illustrated at the bottom of the figure, had a piston at both ends—a variation on the 2-X design under development by Firestone for

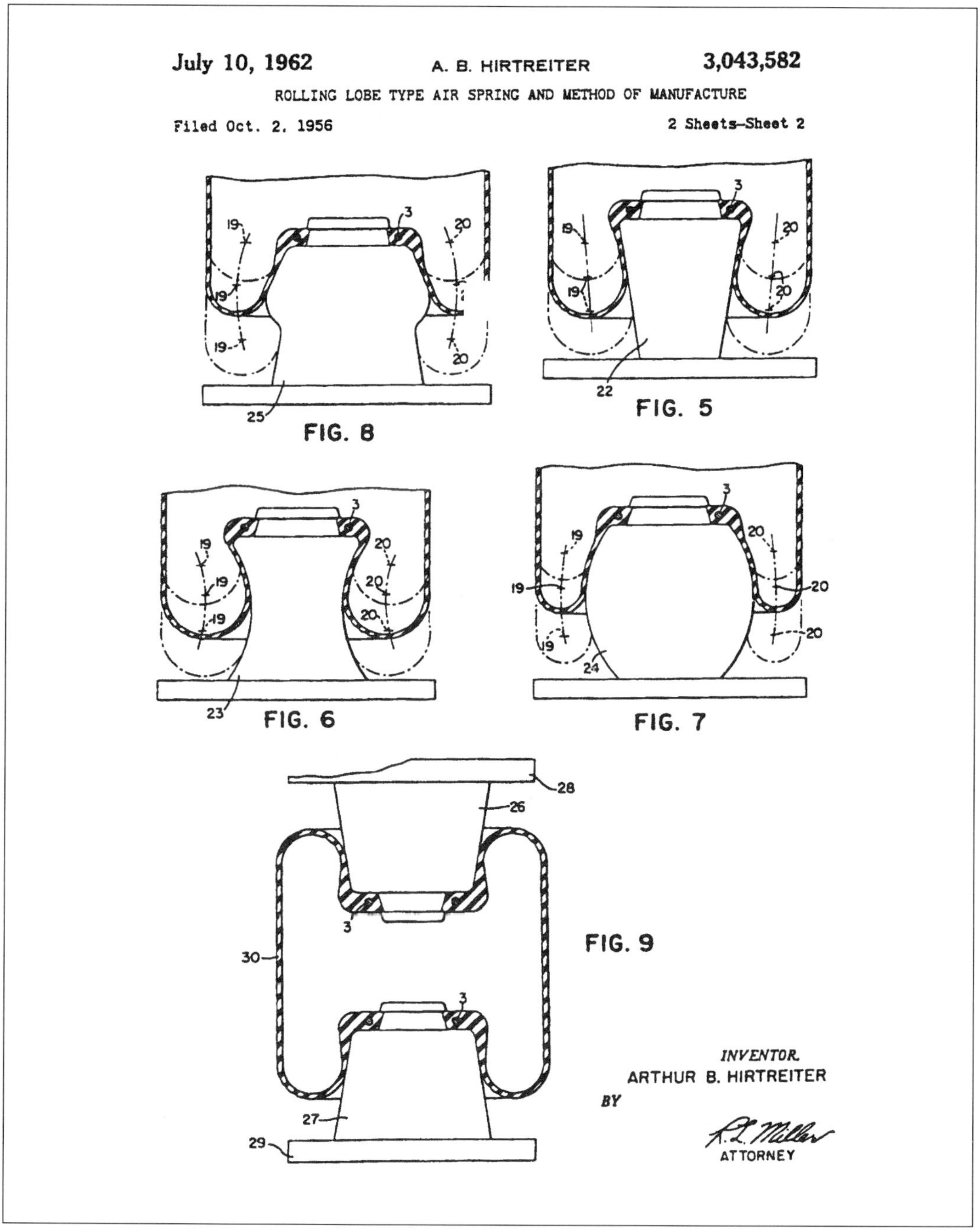

Fig. 7.2 Second page of Arthur Hirtreiter's Goodyear patent, showing a variety of piston shapes and, at the bottom of the page, a configuration that is nearly identical to the 2-X design that Firestone was proposing to Ford at the time of the application (1956)—except it requires no restraining girdle band.

Ford at the time, but without the wide external girdle band. Further analysis revealed, however, that a lower spring rate for a given package size could be achieved with a single piston—in effect utilizing the space vacated as displacement volume.

That the principles involved in quantifying rolling lobe theory were permanently burned into Hirtreiter's memory at the time, is evidenced by an attachment to his letter to me of April 28, 1998, entitled "Elmer Kimmich's Derivation of the Equilibrium Angle for an Unrestrained Fabric Cylinder," a hand-written document in which, at age 84, inventor Hirtreiter managed to reconstruct this mathematical derivation (incidentally apologizing for "how rusty my math was").

Although I will preserve Art's letter, the mathematical details derived therein are beyond the scope of this work—or, frankly, this writer. But Tom Bank's comment on the same subject in his e-mail note to me of April 21,1998, reveals the gist of Art's calculation, and the like-mindedness of these two engineers:

"It is obviously related to the well-known design criterion for pressure vessels," Tom's e-note said, which "states that the lateral forces are twice those of the longitudinal plate forces. This, of course, leads to a cord angle necessary to balance these forces. But," Tom cautioned (at age 80), "in calculating the correct angle you have to take the *stretch* of the cords into account, plus the rubber stresses [between plies], as well as the rubber wall modulus and thickness."

This kind of analytical, intellectually-disciplined approach to product design, still displayed by these two air spring pioneers decades after their retirement, is in refreshing contrast to the didactic manner of some of the inventors whose works are cataloged in earlier chapters. Although there have been many other contributors to the technology, I believe an industry was founded upon the work of Tom Bank and Art Hirtreiter.

Mounted on a Goodyear fatigue test machine in Fig. 7.3 are a pair of rolling lobe air springs of the size and with the straight-sided piston used on the rear of the 1960 GM New Look coaches.[3] Fig. 7.4 shows a slightly smaller size being assembled on a load-deflection test machine. This one has an hour-glass piston of the kind installed on the front coach springs in order to depress the natural frequency.[3] It can be seen that the bottom of the molded part is open to take advantage of the displacement volume of the hollow aluminum piston. In Fig. 7.5, three (inverted) rolling lobe springs of two different sizes are shown undergoing a durability test in Goodyear's laboratory.[4] The top of these assemblies is closed by means of an "upper retainer" and a clamping plate that is drawn up to the chassis by means of integral studs, as seen in cross section in Fig. 7.6.[5]

Fig. 7.3 Rolling lobe air springs having the straight-sided aluminum pistons used on the rear of GM buses, on test in the Goodyear laboratory.

Fig. 7.4 Showing the molded shape of the rolling lobe. The shape of the "hourglass piston" depresses the effective area curve in the middle, near the design position, to achieve a low natural frequency. The bottom of the part is open to utilize the displacement volume of the hollow piston, which also helps to decrease the rate of the part.

Fig. 7.5 Inverted rolling lobe air springs of two different sizes, both with hourglass pistons, on test in the Goodyear laboratory.

With an hourglass piston similar to the one shown in these figures, Hirtreiter was able to achieve a natural frequency of 45 cpm on the springs installed at the front axle of the GM buses. The natural frequency on the larger-diameter rear springs, with a straight-sided piston was 60 cpm. A set of load-deflection curves at three different pressures is presented in Fig. 7.7.[5] Not long after the Goodyear rolling lobe went into production at General Motors Truck & Coach Division—and to Hirtreiter's considerable chagrin—in order to settle a patent dispute in which Goodyear may have been infringing a Firestone tire patent, the law departments of the two companies settled the matter in a trade permitting Firestone to make air springs in accordance with Goodyear's rolling lobe patent in exchange for Firestone's dropping its tire patent protest.

In producing its similar part, Firestone incorporated the positive-seal, clamped bead plate that Tom Bank had developed earlier to close and hermetically seal the top of convoluted springs—a feature later adopted by Goodyear as well. The resulting Firestone assembly is pictured in Figs. 7.8 and 7.9. The steel bead plates were (and are) made with either integral studs as shown in Fig. 7.8, or with recessed blind nuts, as shown in Fig. 7.9. Both have a recessed air fitting (also visible in Fig. 7.9).[6]

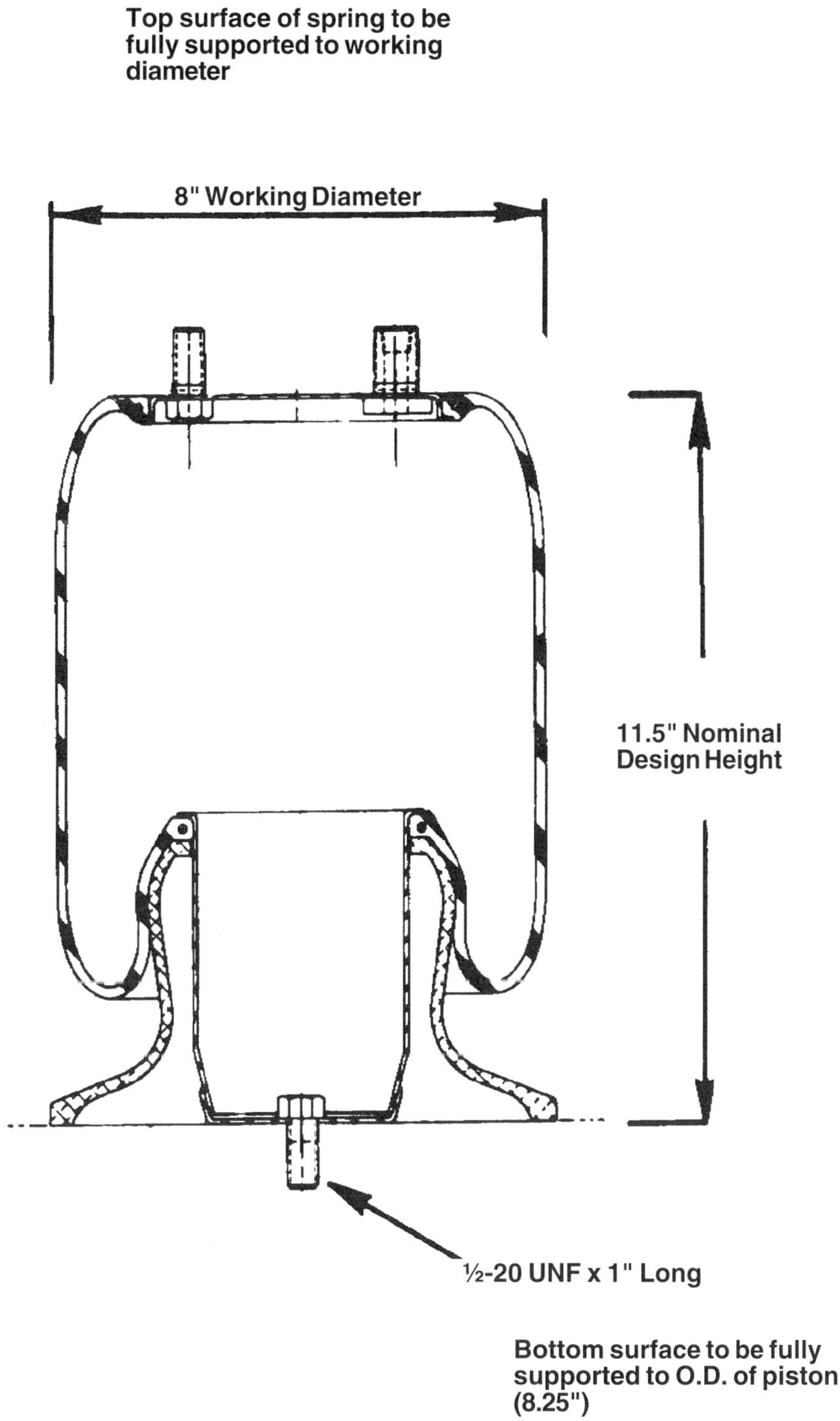

Fig. 7.6 Cross section of the rolling lobe design, showing the upper retainer and clamping plate that sealed the top of the spring.

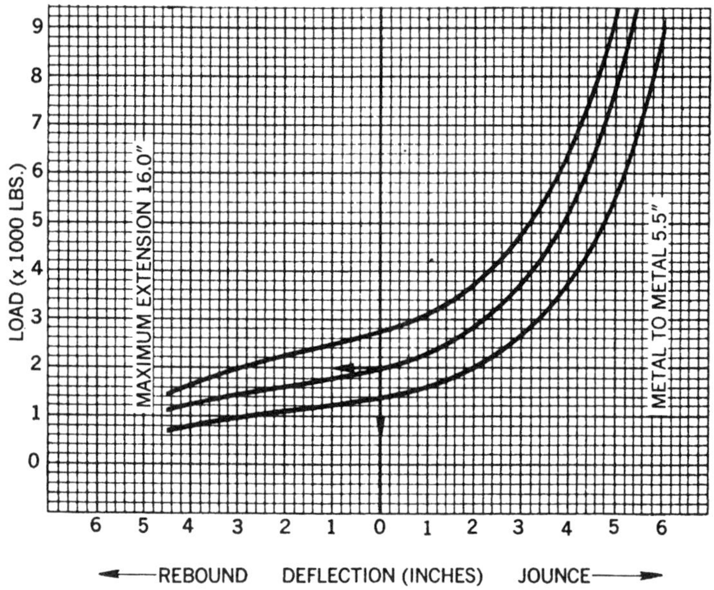

Fig. 7.7 A set of rolling lobe load-deflection curves at three different pressures.

Fig. 7.8 The top of the Firestone rolling lobe, or "rolling sleeve" as it called by that company, was closed with the crimped steel bead plate originally developed by Tom Bank for two-convolution bellows—this one having integral studs and a recessed NPT air fitting. Early Firestone pistons were made of hard rubber, later becoming either aluminum or glass-filled nylon.

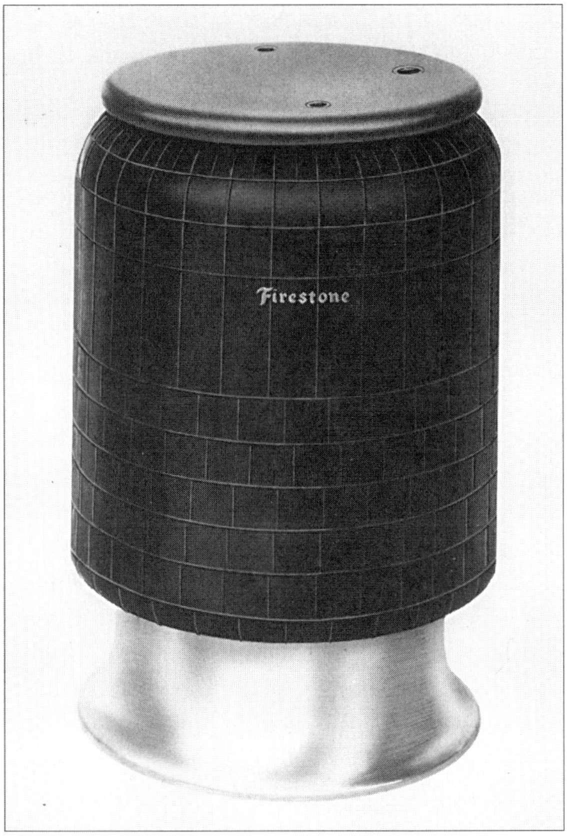

Fig. 7.9 Firestone bead plates are also available with recessed blind nuts, together with an NPT air fitting. This spring has an aluminum piston.

The earliest Firestone pistons (intended initially for either Type **B** rolling diaphragm or, later, Type **D** rolling sleeve springs) were made of hard rubber, which proved to be too brittle for the application. (They broke if they were dropped.) Beginning in 1961, pistons were made of aluminum (Fig. 7.9), and, more recently, of glass-filled nylon (Fig. 7.8). They continue to be produced from both materials today, depending on the size and model of the air spring.

As indicated above, with their new rolling lobe air springs, the natural frequency of GM's New Look transit buses was 45 cpm on the front and 60 on the rear, a dramatic decrease from the 75 cpm front and 80 cpm rear characteristic of the first generation of Airide coaches. But on city streets with uneven pavement, e.g., dips at storm sewers, and especially on curves, the softness of the GM suspension sometimes resulted in excessive roll. On a vehicle with the relatively high center of gravity characteristic of buses, such sideward sway will subject passengers to significant *lateral* accelerations—the mode least tolerated by the inner ear. The problem came to a head at Queen's Transit during the New York World's Fair at Flushing Meadows Park.

"I got a call from John Locklin at GM one day," Art Hirtreiter related. "'Kids are coming home from the fair,' John said, 'where they've taken the bus from the fairgrounds to the parking lot, and they're getting car sick. And they have to clean out the buses every trip, practically.'

"So I went down and, sure enough, what was happening was, as the buses were coming from the fairgrounds going onto the Belt Parkway in a double loop, they leaned to one side and then the other and swayed. They *really* swayed! So I thought we could do something. I got hold of Bill Schmitz at Delco [the GM Division that made the height-control valves for the bus], and I asked Bill how close he could hold the height of the bus with the leveling valves. And the answer was he could hold it within half an inch."

Hirtreiter first used stacks of metal washers in his empirical study to optimize the pistons for the springs in the laboratory, ending up with a new shape having a relatively short straight section, flaring the part toward the base to rapidly increase the effective area, thus snubbing the jounce stroke. He had a set of aluminum pistons made, put them in his car, and drove to Long Island, where Locklin, Killinger, and their boss, Coach Engineer Don Manning were waiting, together with the general manager of Queen's Transit. The group rode the coach and approved the change on the spot. Hirtreiter thinks of his solution, which effectively reduced pitch and roll, as a passive "semi-active suspension."

While waiting for new patterns to be made and production castings to arrive, as a fix for buses in the field, GM resorted to installing blocks of wood under the existing pistons (and lowering the height control valve setting) to engage the flare at the bottom of the existing piston sooner in the jounce stroke.[7]

A decade later, by which time Don Manning had become assistant chief engineer of GM Truck & Coach Division, Manning added the feature of deflating the right front air springs to lower the step height on demand. Fig. 7.10 shows how this feature lowers the bus step nearly to curb height, facilitating boarding for children, the handicapped, or elderly passengers.[8] The inflated step height is shown in the two pictures at the left, in contrast to the step height in the "kneeling" position at the right. The feature was retained on the next generation of buses, the RTS (Fig. 7.11)—the last bus to be made by General Motors, which discontinued coach production in the late 1970s.

By then, the bus industry had changed substantially. Greyhound had begun manufacturing its own coaches at its MCI subsidiary, the transit bus market had declined substantially, and foreign manufacturers, including Neoplan and M.A.N. of Germany, Scania Vabis of Sweden, and Flyer of Canada, had entered the American transit market.

Fig. 7.10 GM buses of the 1960s added the feature of deflating the right front air spring on demand, making the bus "kneel" to facilitate boarding children, the handicapped, or elderly passengers.

Fig. 7.11 In addition to air suspension, when GM's "RTS" transit bus was first released it had a pneumatic "HELP" energy absorbing bumper (a product of the author's Firestone New Product Development Group, which became standard on all new buses, U.S. Postal Service vehicles, and many delivery vans). Refinements in the polyurethane elastomer and internal ribbing later eliminated the need for inflation.

In early 1960, as the major passenger car manufacturers were phasing out air suspension, Art Hirtreiter and his project engineer, Tom Burkley (who would later succeed Hirtreiter as chief engineer), were visited by a group from General Motors' Delco Division in Dayton. The Delco engineers explained that they had been working on a shock absorber that could be pumped up with air to carry additional load when a station wagon, for example, had a heavily-loaded rear end, or when a trailer was hitched to a passenger car.

"They brought in a box of parts that included a shock absorber around which they had assembled one of the rolling lobe springs we had made for American Motors," Burkley explained. "It was a fully molded rolling lobe air spring with beads at both ends, constructed like we made all of them at the time. They wanted to know how much a spring like that, customized for their application was likely to cost.

"I think we said, 'Maybe around five or ten dollars,' and they about dropped through the floor. But we told them we were interested in working with them, and we asked what kind of a price we'd have to shoot for to make it practical for them. And they said 'a dollar!'

"So we first said, 'Wow!' ... But then, 'Maybe we could make this like a piece of hose, and we can work together about how we might clamp the part together.' And for the next year I worked on 141 different constructions to finally get one that would have a flex life of a million cycles."[9]

Both Delco and Goodyear built machines for durability testing of the product—with the usual customer-supplier discrepancies in results. The final successful construction consisted of a thin *extruded* inner liner tube, two plies of fine denier tire cord (60 cord ends per inch [24/cm]) and a wrapped-on cover—the all-Neoprene part having a wall gauge of 0.11 in. [3 mm].

It took two Burkley innovations before Delco's flex-life goals could be achieved. Studying the flex patterns after failure, Burkley conceived the idea of molding circumferential grooves into the liner to reduce the stresses in the rubber on the inside—thus also reducing the stretch in the cover.

"I tried to duplicate the failure pattern we were getting on the inside," Burkley explained. "At first, we tried a pattern of grooves kind of like corduroy, but that pattern would hang up when the part wasn't pressurized. So then I realized we needed a short valley and a flat, and then another shallow valley, and then, if the pattern was on a helix, the flats could slide over each other. But we still had a problem with chafing when the shock wasn't pressurized. So we borrowed from our run-flat tire technology, where we used liquid silicone to lubricate the inside of the part so it wouldn't chafe—and that did it."

The helical pattern is machined into the mandrel on which the part is built and cured, as seen in Fig. 7.12. A double-length pair of sleeves are built on each mandrel and later cut apart. The mandrel is inserted into one of the five barrels of a horizontal Goodyear curing machine nicknamed the "Gattling Gun" (Fig. 7.13), a Burkley design. An external curing bladder, initially withdrawn by vacuum, snaps down around the mandrel to apply pressure during cure. Fig. 7.14 shows several of the resulting clamped sleeves on durability cycling test in Goodyear's laboratory.

Tom Burkley's innovation became U.S. Patent 3,691,268, dated September 12, 1972, as seen in Fig. 7.15. A prolific inventor, Burkley received a number of Goodyear awards for his contributions over the years. In 1994 he received the Distinguished Corporate Inventor of the Year Crystal Award from Akron's Inventure Place (the Inventors Hall of Fame).

Fig. 7.12 Mandrels on which Goodyear built Delco clamped reversible sleeve ***Superlift*** *air springs had a helical pattern to improve durability. The extruded inner liner was snapped around the mandrel before fine denier tire cord plies were wrapped on.*

Fig. 7.13 Goodyear's "Gattling Gun" curing machine into which the mandrels were inserted for vulcanization.

Fig. 7.14 Clamped Delco air springs, frequently called "shock sleeves" because of their application, on durability test in the Goodyear laboratory.

United States Patent [15] **3,691,268**
Barkley [45] **Sept. 12, 1972**

[54] **METHOD OF MAKING A FLEXIBLE MEMBER HAVING INCREASED SERVICE LIFE**

[72] Inventor: **Thomas E. Burkley**, Akron, Ohio

[73] Assignee: **The Goodyear Tire & Rubber Company**, Akron, Ohio

[22] Filed: **March 23, 1970**

[21] Appl. No.: **24,931**

Related U.S. Application Data

[62] Division of Ser. No. 679,731, Nov. 1, 1967, Pat. No. 3,598,155.

[52] **U.S. Cl.**.................**264/162**, 264/226, 29/157 R, 51/290

[51] **Int. Cl.**...........................**B29c 1/02**, B29h 21/08

[58] **Field of Search**......264/225, 226, 227, 219, 220, 264/162, DIG. 40; 18/47 R, DIG. 12; 29/157 R; 51/290

[56] **References Cited**

UNITED STATES PATENTS

1,951,377	3/1934	Twiss et al.	264/219 X
2,288,840	7/1942	Raiche	18/47 R X
2,958,148	11/1960	Sylvester et al.	264/220 X
3,313,873	4/1967	Dembiak	264/219 X
3,473,989	10/1969	Richmond	264/227 X

Primary Examiner—Robert F. White
Assistant Examiner—Allen M. Sokal
Attorney—F. W. Brunner and Robert H. Hatton

[57] **ABSTRACT**

A method of making a flexible member having increased service life with the member being of the type designed for use in a fluid pressure system in which a predetermined flex wrinkle pattern or a modification thereof is imparted to a surface of the flexible member to improve the flexing properties while the member is operating in such system. This pattern may consist of a simulation of the actual flex wrinkle pattern of a previously operating flexible member or be a compromise which employs a plurality of grooves having substantially the same pitch as the wrinkles. This invention results in the increased flex life and improved low temperature performance of the flexible member.

12 Claims, 8 Drawing Figures

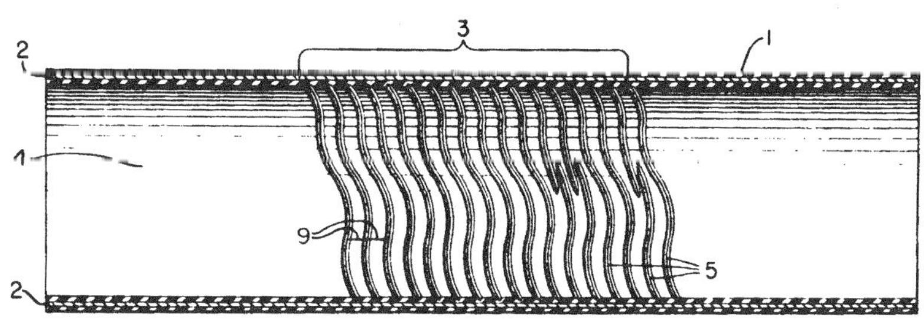

*Fig. 7.15 Goodyear engineer Tom Burkley's patent for his methods of manufacturing the small-diameter clamped Type **D** air springs.*

Delco's finished product utilizing the Goodyear sleeve is shown in Fig. 7.16, as advertised in an early Delco brochure. The part is seen in cross section in Fig. 7.17.[10] Intended for the aftermarket, the *Superlifts* on the loaded car were designed to be inflated at a service station, pumping air in until the car was level. One of the two fittings seen in the figure was attached to a hose which terminated in a filler valve at a convenient location on the car exterior, as seen in Fig. 7.18.[10] The other fitting was attached to a "crossover tube" which equalized the pressure between the two rear shock absorbers, also making it unnecessary to inflate the two shocks individually. By 1963, Goodyear was making more than 80,000 beadless rolling sleeves per month for Delco's *Superlift* shock absorber.

The Delco shock absorber got the attention of engineers in other GM divisions. By 1962, although it had discontinued air suspension at the end of 1960, Cadillac Engineering had developed a fully automatic system, integrating the *Superlifts* into the rear coil-spring suspension of its cars. Initially, the system used a small vacuum-driven compressor in the engine compartment, but this unit was soon replaced by an electrically-driven compressor which was turned on as needed by an electric height-control switch. The system remains standard on Cadillacs today.

Soon after its development, the Air Lift system was also offered on the El Camino pickup truck, made by Chevrolet, and it has since become either standard or optional on models made by Buick, Oldsmobile, and Pontiac. It is seen on the rear of a 1998 Cadillac Seville STS in Fig. 7.19.[11]

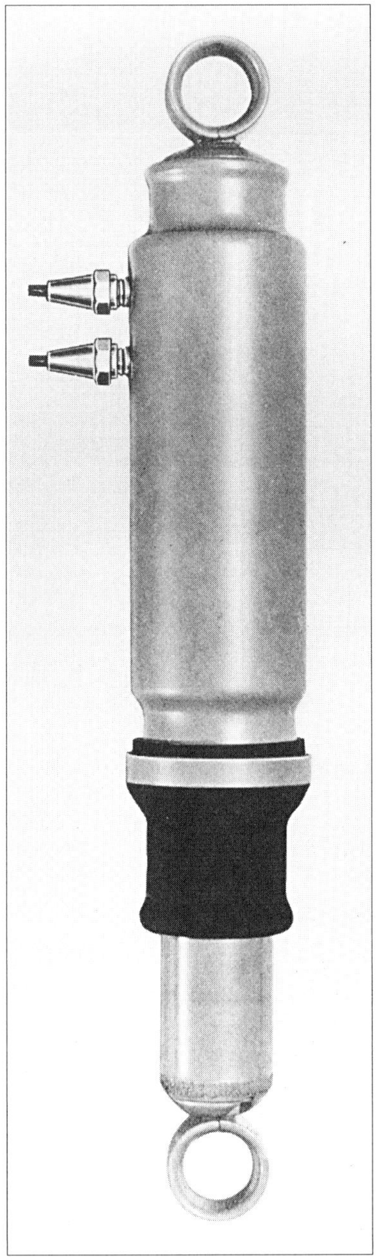

*Fig. 7.16 Cover of a vintage Delco brochure introducing the **Superlift**, a combination shock absorber and air spring for the automotive aftermarket.*

While the Delco sleeve was under development, the Gabriel Shock Absorber Company of Cleveland had been working on a similar idea, except that its sleeve was an internal member filled with oil inside a shock absorber. The sleeve rolled between the telescoping tubes—somewhat in the manner of a hydropneumatic

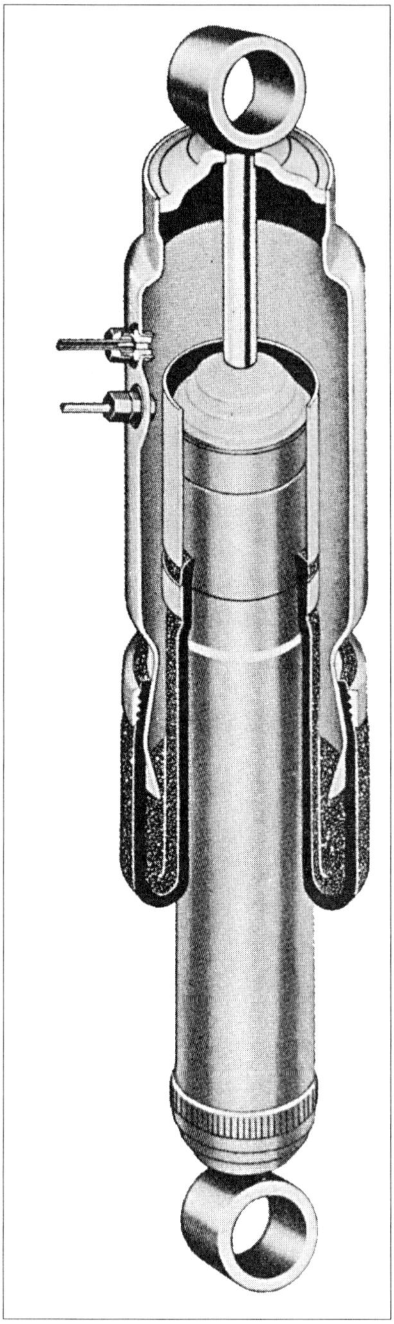

Fig. 7.17 Cross section of the Delco **Superlift**, showing how the air spring uses the shock absorber exterior as its piston.

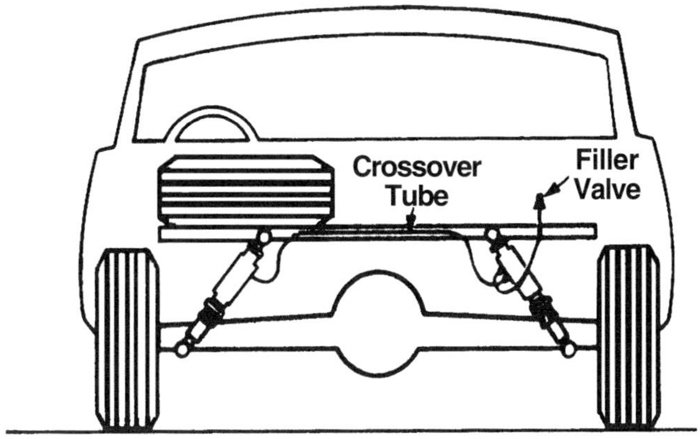

*Fig. 7.18 Diagram from the vintage Delco brochure, showing how **Superlifts** were installed on the rear suspension of the car. One of air fittings on the product was attached to a hose terminating in a Schraeder valve to pressurize the pair; the other fitting received a hose connected to the second air spring.*

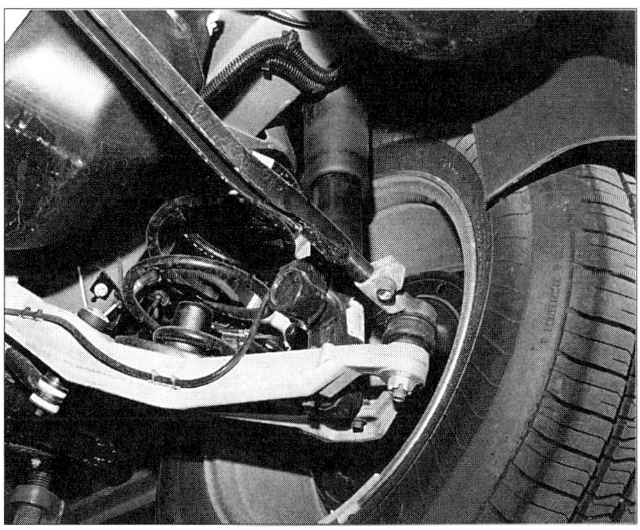

Fig. 7.19 Standard equipment since the early 1960s, the Cadillac Air Lift on the rear suspension of a 1998 Cadillac Seville STS. (Photo by the author, courtesy of Dave Towell Cadillac, Akron.)

spring (see Chapter 9). The company that had initially been making the reinforced rubber part had been constructing it as a single-ply butt-spliced sleeve. The cords were parallel to the axis (at what an air spring builder would call zero-degree bias angle). Having no circumferential strength in the meniscus below the fully-supported cylinder exterior, Gabriel's early units frequently failed between the parallel cords, leaking oil and staining driveways.

Gabriel eventually came to Goodyear. Using a very fine (60 ends/inch, 210-2 denier) cord fabric, Hirtreiter and Burkley developed a low cord angle (10° opposing angles) Neoprene reversible sleeve which worked inside the shock absorber, providing Gabriel with a salable unit.

By 1971, Monroe Auto Equipment of Monroe, Michigan had entered the market with a product very similar to the Delco *Superlift*. Dubbed the *Ride-Leveler*, the Monroe "air adjustable shock absorber," used clamped exterior air spring sleeves made by Goodyear. By the end of the decade there were a number of competitors for the air-adjustable shock absorber market, including a Chicago company whose product was called *Weightlifters*. Vintage brochures from these companies are pictured in Fig. 7.20.[12]

Fig. 7.20 A collection of vintage brochures from the 1960s and '70s showing competitive products in the air-adjustable shock absorber market.

The ability to make low-cost, small-diameter, durable air springs simply by clamping rubber-fabric sleeves to adjacent metals opened more than one new market for Goodyear (and, in due course, for Firestone). One of these became pneumatic air suspension for *truck seats*.

As is apparent from the rather high natural frequencies selected for the new General Motors trucks described in the next chapter, the initial movement to air suspension on commercial vehicles was focused primarily on decreasing vehicle maintenance, an advantage that had been documented on buses. Road shocks transmitted into the structure, especially under partial loading conditions, could be dramatically reduced. One had always expected these vehicles to "ride like a truck."

Driver comfort as late as the 1950s was a secondary consideration. But as fleets competed for professional transcontinental drivers and the Teamsters Union gained strength—and more drivers began to own their own rigs—driver comfort became a much more important factor.

It was Kenworth Motor Truck Company, the premiere Seattle truck manufacturer who had already adopted air suspension for its tractor suspension (using a larger-diameter version of the Ford **C**-type 2-X spring developed by Charles Slemmons), who, working with Goodyear, introduced the Kenworth Air Seat. As advertised in its 1961 brochure (Fig. 7.21), this 47 lb [21 kg] fiberglass and aluminum seat rested on a single air sleeve, and it had a small shock absorber to control road-induced bounce.

The popular Bostrom Seat (Fig. 7.22) followed. The air sleeve can be seen in the figure. Fig. 7.23 is a modern seat spring for a National Seating Company *Cush-Naire* seat. Most of these designs, e.g., the *Air Command* (Fig. 7.24), have a scissors arrangement which reduces the ride frequency of the seat to less than half the natural frequency of the air spring.

This would not be the end of applications for clamped air sleeve technology. In another decade *cab mountings* for truck tractors would be produced from air sleeves, a development that was reported in an SAE technical paper in 1978 by Albert W. Foster, who describes a sophisticated air suspension subsystem developed by Ford Motor Co. Pivoted at the front on rubber bushings, the cab system has its own height-control valves and shock absorbers that limit resonant interaction with the primary suspension system of the truck. The natural frequency of the Ford cab system was in the 1.0 to 1.5 Hz [60–90 cpm] range, but this proved sufficient to substantially reduce the vertical transmissibility of

New!

KENWORTH AIR SEAT

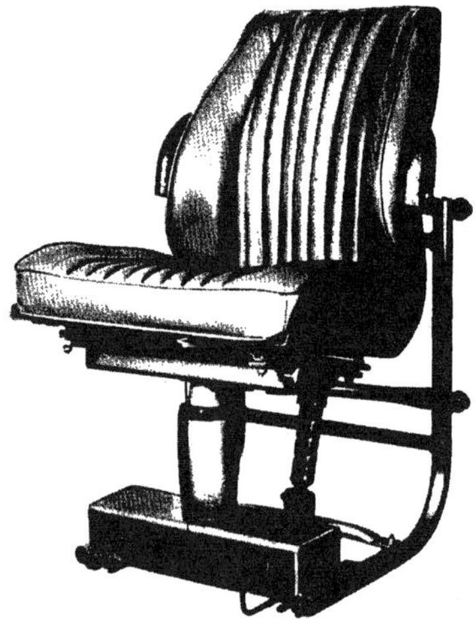

Now – a completely new driver's seat that will improve the riding qualities of any heavy-duty truck.

The new Kenworth Air Seat, designed to provide comfort and maximum vehicle-control for the driver, has been tested under the most rugged conditions –– on off-highway road construction equipment.

Outstanding features of the Kenworth Air Seat include these:

- Rides on air – supported by an oversized air cylinder.
- Completely and easily adjustable for height, fore and aft position and reclining angle.
- Contoured shell of heavy fiber glass.
- Air foam cushions, covered with vinyl upholstry used in fine automobiles, custom hand-crafted and pleated.
- Riding action of seat is vertical – no pitch or sway.
- Direct-acting, long-life shock and rebound absorber.
- Needs no lubrication or maintenance – all joints are rubber in compression.
- Frame parts are built of light-weight aluminum.
- Lightweight – entire seat weighs only 47 pounds.

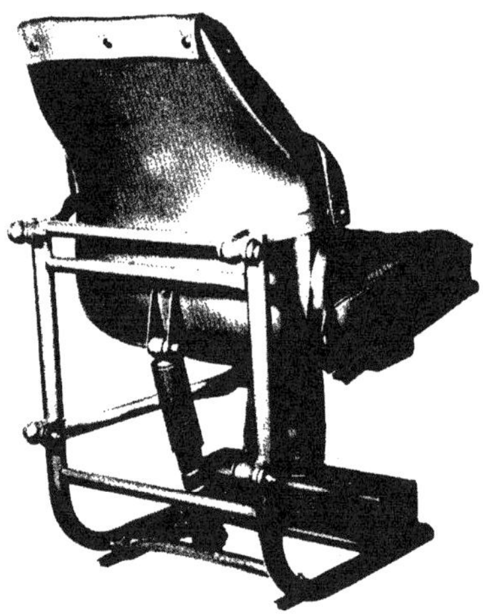

For real driver comfort and reduced driver fatigue, see your nearest Kenworth distributor now about the Kenworth Air Seat – another "first" in Kenworth's long list, proving once again that. . .

There's more WORTH in KENWORTH

Patents Pending SP–11–24–61

KENWORTH MOTOR TRUCK COMPANY

HOME OFFICE: SEATTLE 24, U.S.A.; FACTORIES AT SEATTLE AND VANCOUVER, B.C.; DISTRIBUTORS IN THE UNITED STATES AND MANY FOREIGN COUNTRIES

Fig. 7.21 Kenworth was the first to put air sleeves into the new application of ***air-suspended truck seats.***

Fig. 7.22 The popular Bostrom truck seat, showing the air sleeve inside the suspension linkage.

Fig. 7.23 The air spring employed in National Seating Company's Cush-Naire seat.

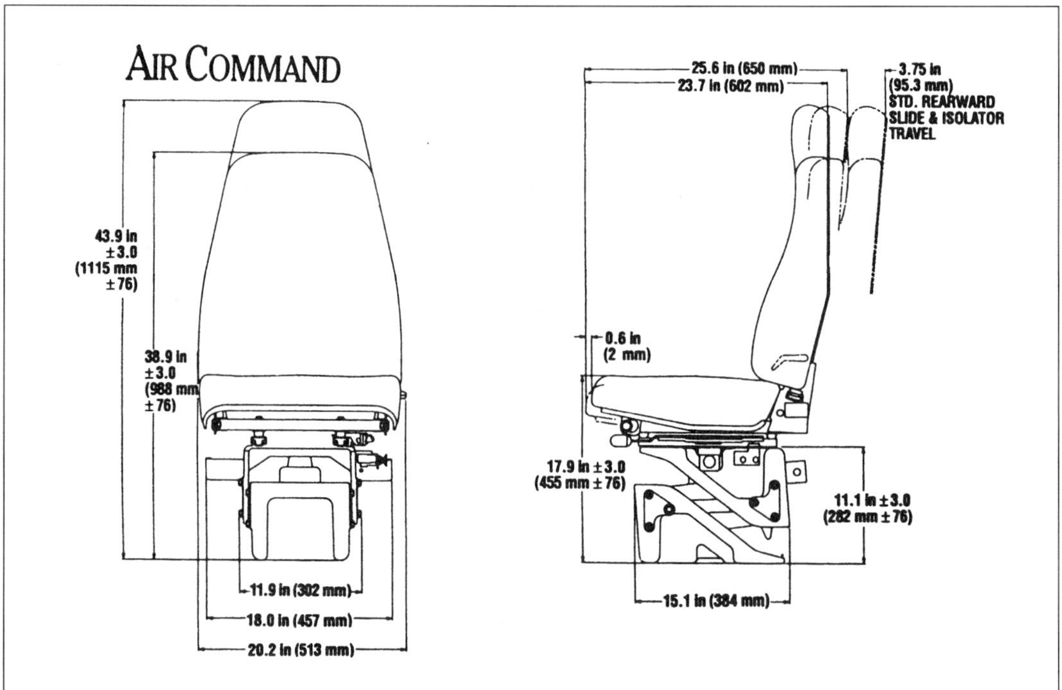

Fig. 7.24 The ***Air Command*** *seat, showing the scissors linkage typical of air seat suspensions.*

frequencies in the 3–7 Hz range, which, the author explains, is the range that produces the most human discomfort. These values were ascertained in subjective sensitivity experiments during which accelerometers were actually taped, just below the navel, to the skin of truck drivers participating in the study.[13]

Finally, in a completely different kind of application for clamped air sleeve technology, Ohio inventor Fred Solomon began putting traditional *farm windmills* back to work in 1996. Overcoming the friction and other problems associated with the sliding leather cup seals used for more than a century in these primitive pumps, Solomon's *HydroGlide* advertises a 40 percent improvement in pump efficiency with his rolling seals—an additional 120 gallons of water per hour. As this is written, two years later, the inventor has perfected a solar-powered freon-driven motor to operate his frictionless high-efficiency rolling lobe pump.

Chapter Eight

PROLIFERATION

Meanwhile, back at the ranch—or on the Indiana prairie—in my new assignment as Airide Product Manager at Firestone Industrial Products Division's headquarters in Noblesville, I had accepted a career gamble. With Firestone's having lost both the passenger car air spring business and, at least temporarily, the GM bus business, my department and I faced the "challenge" of producing a million dollars in air spring sales between January, 1960, and the end of the fiscal year on October 31st that year. The Division's newly built air spring plant, constructed in anticipation of huge passenger car volume, was, at the moment, contributing spectacular *losses* to Firestone's monthly P & L statement. If we did not achieve our million dollar goal (which, on paper, represented a potential break-even point), the company would go out of the air spring business, and the future employment of the small Airide staff, including this member, would be in doubt.

Art Hirtreiter, and Goodyear's air spring sales manager, Andy Palguta, weren't much better off. Art's department had by now shrunk to three, including himself. Having lost the Ford business, Charlie Slemmons was in similar straits at General Tire. U.S. Rubber had already abandoned the product. The future of air springs was less than bright.

Firestone Industrial Products Company already had a staff of about two dozen salesmen, but they sold no proprietary Firestone products. Their job was to call on purchasing agents, requesting the opportunity to bid on any industrial rubber parts their customers' engineering and production departments had designed and put out for purchase—components that would become a part of the industrial or consumer products that these companies manufactured. The salesmen picked up blueprints of washing machine wringer rolls and drain hose, private-brand lawn mower and tricycle tires, automotive engine mountings, axle bumpers, accelerator pedals, radiator hose, foam rubber seat cushions, and other custom parts—for which, if Firestone got the order, customer-owned molds would be made.

It was unreasonable to expect these sales personnel to sell air springs at a purchasing level. Most purchasing agents didn't know what an air spring was (having never received a purchase requisition to buy any). Indeed, most engineers in non-automotive companies were not much better informed about the product. A need for our air springs obviously had to be developed at an engineering level. To this end, I had a staff of three field engineers: Mike Kray, Jerry Marsh, and Bob Weir—the last of whom would retire a few months after I arrived, to be replaced by a recent Georgia Tech graduate, Dan Morris.

We would not be starting from scratch. Although the half million dollars in sales of bus bellows to General Motors had evaporated with the advent of Goodyear's rolling lobe, there were the non-automotive air spring applications developed by Kray, Marsh, Weir, and, earlier, Allen Boyer. And we were receiving some modest orders for two-convolution bellows from Fruehauf trailer, Ole Granning, Mack Truck, Brown Trailer—and there were GM and Mack service requirements. But as things stood, the business we could count on was less than half of what we needed to keep the ship afloat after the end of October.

There were a couple of bright spots. General Motors Truck & Coach Division had recently brought out two new trucks on Firestone Airide springs. The first of these had been released in 1956, but its volume had been less than spectacular. A schematic of the front and rear suspensions of this truck appears in Fig. 8.1. A descendant of GM's earlier development work on buses, both the front and rear suspensions of the truck employed two-convolution bellows with displacement reservoirs to reduce their spring rate. Situated *beneath* the bellows on the rear suspension, these hollow, rectangular reservoirs doubled as tracking members, with rubber bushed ends. On the front suspension, metal boxes, mounted inboard of the bellows, constituted the displacement reservoirs. The system had two leveling valves in the rear and one in front—although, during development, project engineers, in an effort to compensate for frame flexibility, had gone through the fruitless exercise of trying four height-control valves before rediscovering Euclid's declaration that *three* points determine a plane; and that four leveling valves tended to "fight."

To our dismay in the Airide Division, a later version of this truck, the GM *Astro Aire*, eliminated the displacement reservoirs, deliberately stiffening the ride in order to improve roll resistance of the vehicle. This decision, in combination with an exceptionally large jounce bumper mounted *inside* the bellows (further reducing displacement volume), resulted in natural frequencies that, by modern standards, would be unacceptable. As Robert F. Schmidt wrote in his introductory technical paper, apparently in an effort to put a positive spin on the values quoted:

"Ride frequency remains essentially constant from light to loaded (197 cpm to 138 cpm) whereas a steel spring suspension has a wide variation (655 cpm to 179 cpm)."[1] Surprisingly, these embarrassingly high frequencies compared favorably with conventional leaf spring suspensions. Road shock transmitted to the frame was greatly reduced—as is apparent from the "B" curves in Fig. 8.2, which is a set of accelerometer traces

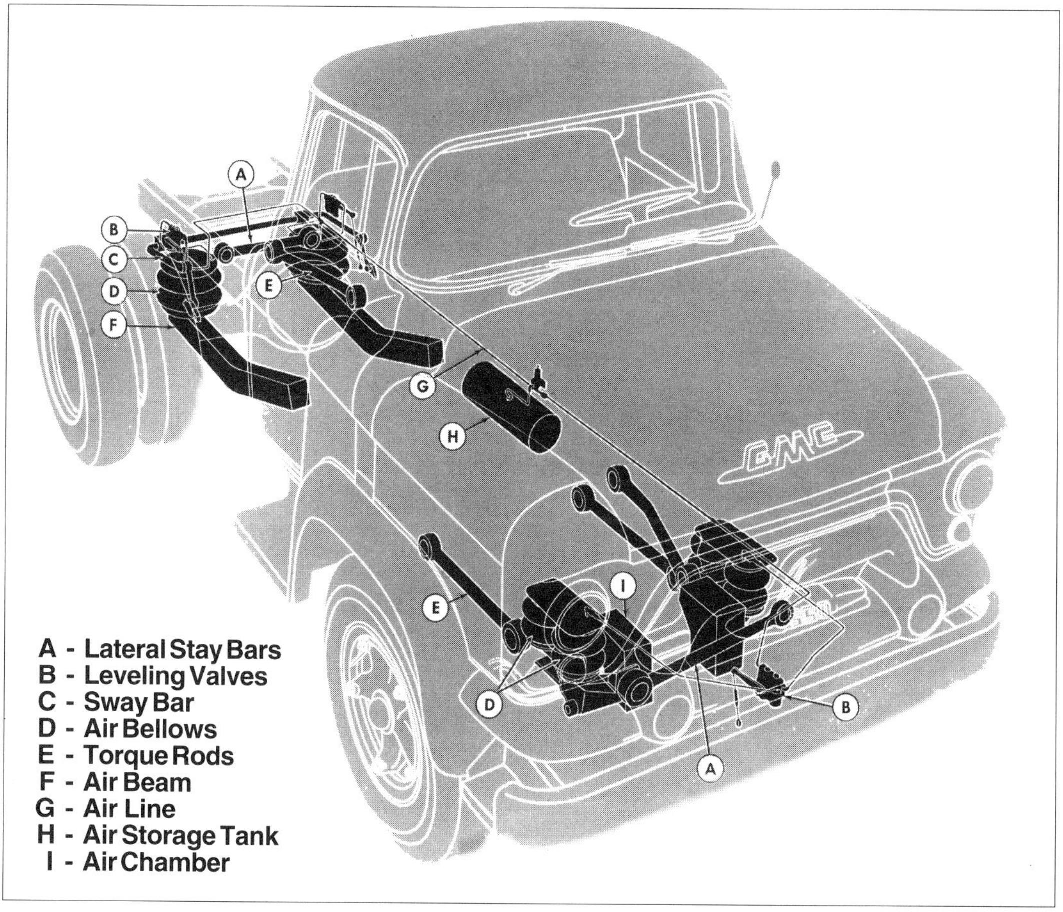

Fig. 8.1 Layout of the GM ***Astro Aire*** *truck air suspension system.*

recorded above the fifth wheel.[2] Even at the natural frequencies produced with two-convolution bellows having no extra volume, over-the-road "jiggling" high-frequency vibrations are dramatically reduced.

Equally remarkable is the reduction in *stresses* measured on *unsprung* members of the suspension. Fig. 8.3 compares the traces of strain gauges mounted on the axles of leaf spring vs. air spring tractors.[2] As GM's Don LaBelle perceptively pointed out in his later SAE technical paper,[3] "it follows that the reactive force against the *road* is significantly reduced." Convinced through statistical analysis of traffic on their roads that air suspension on trucks actually does reduce highway maintenance, European authorities would reduce the taxation on air suspension vehicles two decades later.

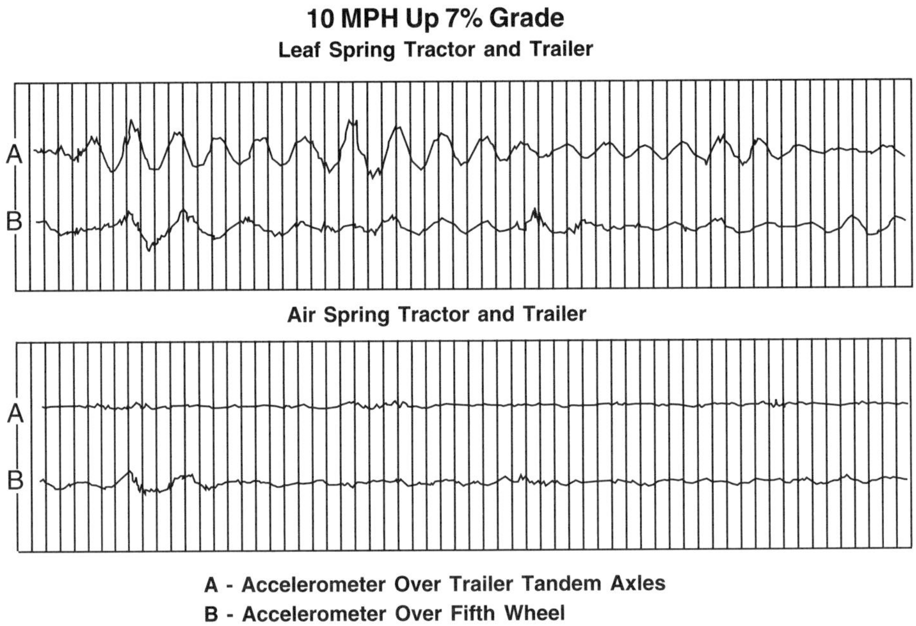

Fig. 8.2 Frame-mounted accelerometer traces compare the shock input of GM ***Astro Aire*** *trucks with and without air suspension.*

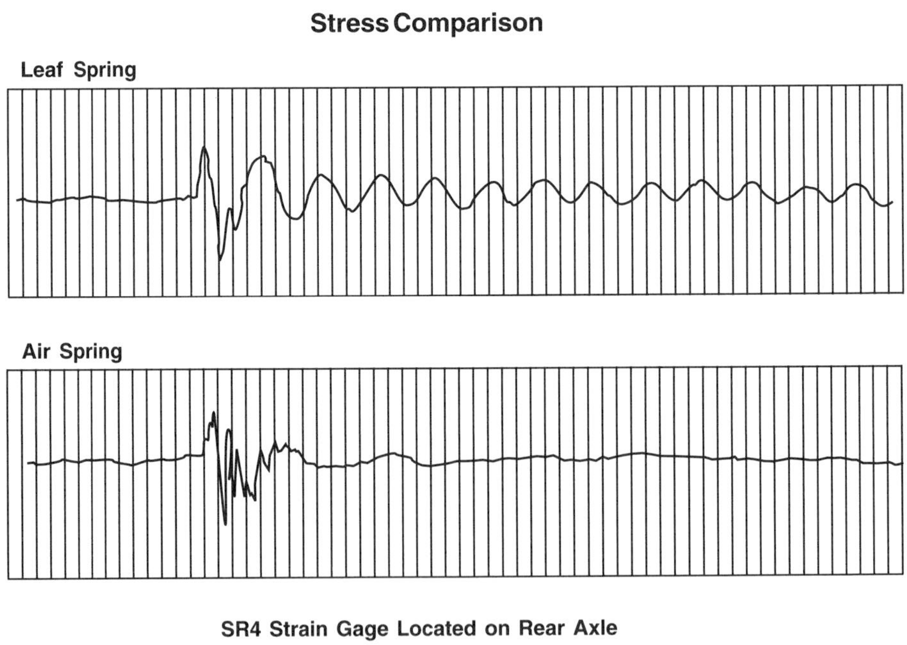

Fig. 8.3 Comparison of traces with an accelerometer mounted on the unsprung axle of GMC ***Astro Aire*** *trucks with and without air suspension.*

By 1959, General Motors had released an all-new, state-of-the-art truck tractor model with a host of innovations. Announced design goals for the model included capacity to carry more cargo weight, the ability to haul a bulkier load, decreased operating cost, and greater driver comfort and safety. The unique profile of the resulting 8000 series is apparent in Fig. 8.4—a radical new concept with a truncated cab which permitted a longer, bulkier trailer. Constructed of riveted aluminum panels, the cab was only four feet long, yet it provided adequate leg room and, with an interior height of 58.25 in. [1.48 m], plenty of head room as well. Without glass and controls, the cab weighed only 280 lb [127 kg].

The vehicle had an exceptionally lightweight frame for a truck. The entire truck tractor, "wet," weighed 10,000 lb [4545 kg]. The tractor, it was claimed, could haul 4000 lb [1818 kg] more payload than conventional highway trucks, partly because of its decreased weight, but also because 2000 lb [909 kg] had been transferred to the front wheels, permitting additional load capacity to be imposed on the fifth wheel. The truck had an independent front suspension and steering linkage, new for trucks at the time. Without air suspension, the ride on such a light tractor would have been intolerable at light loads, even worse when bob-tailing.[3]

In their SAE technical paper, "A New Concept of Light Weight Highway Tractor Design," C. V. Crockett and D. J. LaBelle disclosed a hitherto unexploited, relatively obscure advantage of air suspension: With constant ride height, they explain (eliminating the need to provide for static spring deflection) some 70 cubic feet [2 cubic meters] could be added to the interior volume of a 35 foot [21.3 m] van trailer. As the paper pointed out, air springs on buses had already demonstrated a reduction in damage to the vehicle structure on lightly loaded vehicles. Air suspension would also, of course, fulfill the objective of increased driver comfort.[3]

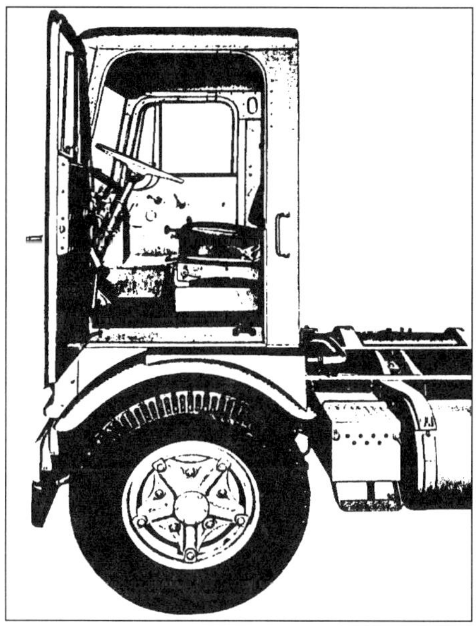

Fig. 8.4 Showing a silhouette of the weight-saving truncated cab of the 1959 GM DLR and DSR-8000. Even with the radical configuration shown, the truck managed to provide adequate interior room for the driver.

The 1-X (**B**-type) air spring configuration (Fig. 8.5) on the DSR-8000 (and its companion DLR-8000), on both the front and rear suspensions, was a larger-diameter version of the product General Motors had developed in cooperation with Firestone for its passenger car line, beginning with Chevrolet. Fig. 8.6 details the effective area characteristic of the air spring, its resulting load-deflection curve, and its spring rate in lb/in. Both the front and rear suspensions of the new truck had a natural frequency of 90 cpm.

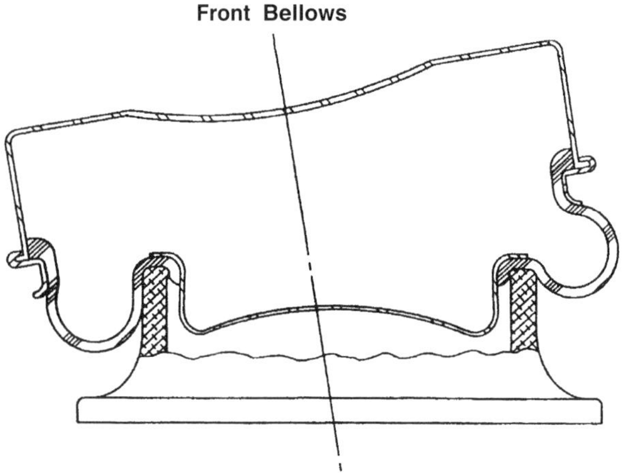

*Fig. 8.5 The large-diameter Firestone 1-X, Type **B** air spring used on the 1959 GM DLR and DSR-8000.*

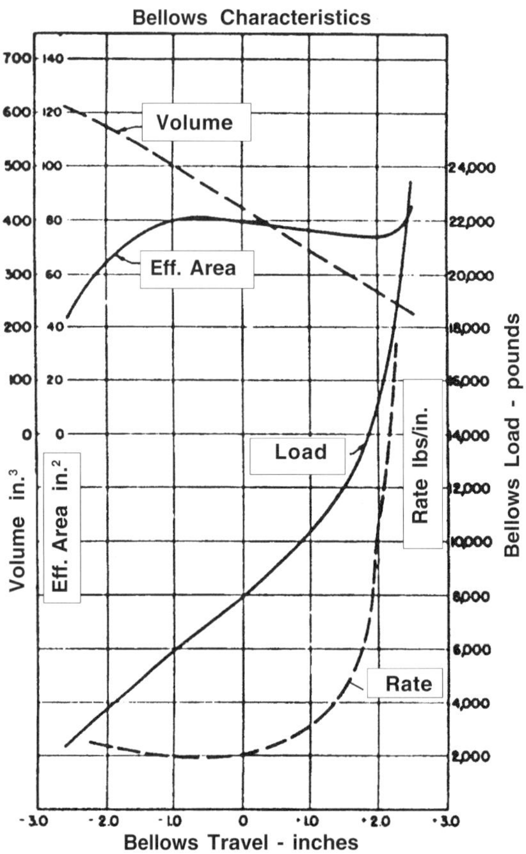

Fig. 8.6 Effective area, volume, load-deflection, and rate curves for the GM DLR/DSR-8000 air suspension.

By 1958, a number of suspension manufacturers were offering pneumatic suspensions for both trucks and trailers. Fig. 8.7 is an early truck drive axle. As Ed Hendrickson, President of Hendrickson Manufacturing Co. pointed out in his 1969 SAE technical paper on the history of tandem truck suspensions, "In most cases the air spring was used as the means of *articulation*, in addition to providing the springing medium."[4] With its spring situated *between* the axles, the early Hendrickson design was a good example.

Several companies, including Hendrickson, Spencer-Stafford, Peterbilt, and Truktor, began with Charlie Slemmons's elongated "Air Pillow" made by General Tire—one long double-convolution spring spanning the pair of tandem axles. On some of these suspensions, this early General spring was installed in place of much rougher-riding rubber-in-shear or rubber-in-compression elements—another innovation that had been pioneered by Hendrickson, and later employed by Utility Trailer and by Chalmers Suspensions.

Fig. 8.7 An early Hendrickson suspension for (tandem) truck drive axles.

Perhaps for the advertising value of appearing to be on the "air" band wagon, Chalmers later introduced a hybrid *Air/Rubber Truck Tandem* putting a single-convolution bellows-type air spring in parallel with its rubber spring. The Chalmers hybrid had no shock absorbers, relying instead on the hysteresis inherent in the large, hollow-cylinder, molded rubber spring. The product found a niche on heavy-duty off-road trucks, including concrete mixers.

Fruehauf developed an air suspension system for its trailers utilizing Firestone two-convolution springs (Fig. 8.8)—a suspension it then offered to smaller trailer manufacturers. Ole Granning and Page & Page were other early suspension makers that employed two-convolution bellows—Page & Page being one of the first to combine an air spring with a leaf spring. None of these had auxiliary reservoirs to depress the natural frequency.

Although Brown Trailer, a division of Clark Equipment, initially had a suspension that used two-convolution bellows, when Clark Equipment developed its own new, *Clark Air Ride* in 1958, the suspension package used an air spring that was a later Slemmons development, a larger-diameter version of the 2-X or **C**-type spring with a wide girdle

Fig. 8.8 The 1960s Fruehauf air suspension employed Firestone two-convolution bellows. The company offered its suspension package to smaller trailer manufacturers.

band that General had made for Ford and Edsel. This same **C**-type air spring was also used on the *Krause Air Ride* suspension. And in an original equipment coup, Slemmons persuaded Kenworth Truck to adopt General's product as standard equipment on the Seattle manufacturer's prestigious Class 8 highway tractor.

Then, in 1960, in a coup for our Firestone group, Neway Equipment Co. of Muskegon, Michigan (one of Gerry Marsh's accounts) completed the development of its *Add-an-Axle* product line. The design made it possible to install one or more independent equalizing air suspension axles on existing vehicles. Before the year was out we would be receiving thousand-piece orders for these Neway air springs, and Tom Bank would begin modifying more of Herb Deist's high-volume passenger car presses to accommodate these large (12.6 inch [32 cm] diameter) parts.

Neway offered suspensions designed for both trailers and trucks. Fig. 8.9, from Lloyd Masser's 1969 SAE technical paper, shows the basic trailing arm configuration of nearly all Neway air axles of the time. Part of the company's successful marketing effort,

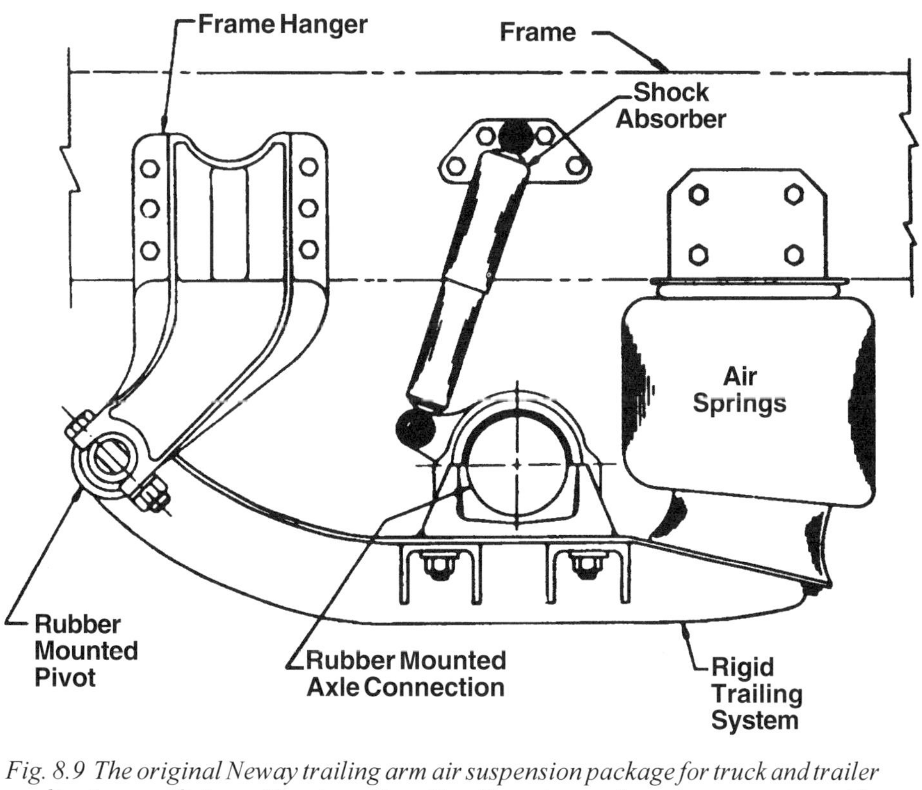

*Fig. 8.9 The original Neway trailing arm air suspension package for truck and trailer applications, utilizing a Firestone Type **D** rolling sleeve air spring, as presented in an SAE paper by Lloyd Masser.*

Masser's was the first of six SAE papers that Neway would publish in subsequent years, including four after the company was purchased by Lear Siegler. Fig. 8.10 shows how an assembly could be mounted as a pusher axle on a tractor; and Fig. 8.11 how additional axles might be added to carry larger loads on a concrete mixer truck.[5]

Neway Equipment later introduced a suspension for tractor *drive axles*, which was a bigger challenge. As Masser's paper points out, "a dead axle air suspension is inherently stable." So in order to provide stability for its "Drive Axle Independent Equalizing Air Suspension," his group devised a parallelogram system to minimize pinion angle change and spline travel during axle articulation.[5]

For those trucks or trailers on which it would be advantageous to lift one or more axles to avoid tire wear when the vehicle was lightly loaded (e.g., on a cement truck), Neway later provided a second (two-convolution) *actuator* air spring on a seesaw arrangement to lift the axle off the road pneumatically, as seen in Fig. 8.12. This device eliminated the unpleasant, time-consuming chore of having to physically chain up an idle axle; and, unlike a steel spring, the exhausted air spring readily collapsed when the axle was elevated.

Fig. 8.10 Showing how the Neway suspension could be mounted on a pusher axle for a truck tractor.

Fig. 8.11 Advertising its "Add-an-Axle" concept, Neway illustrated how additional independent equalizing axles could be installed to permit hauling greater loads without exceeding load limits.

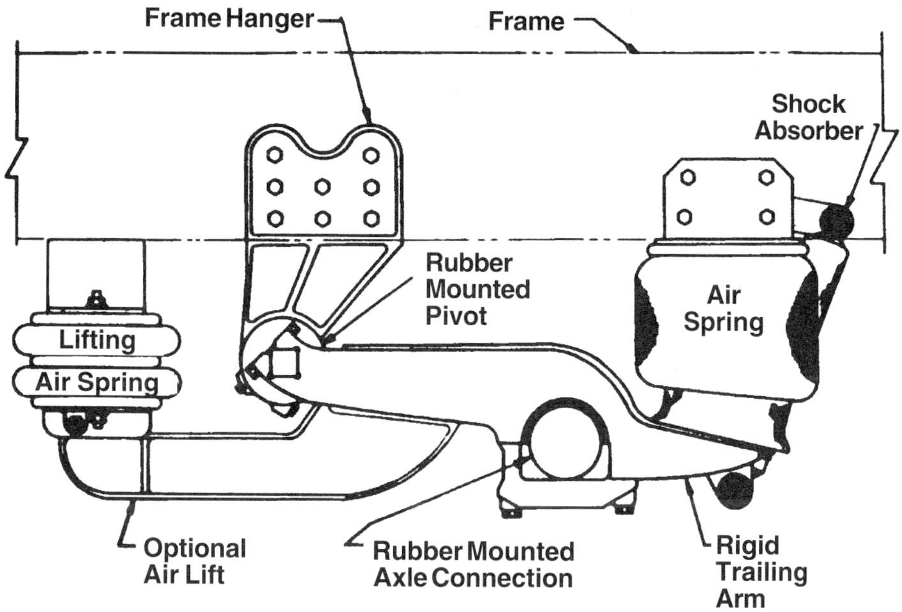

Fig. 8.12 To avoid excessive tire wear when operating with lighter loads, this seesaw geometry lifted the axle off the road when it wasn't needed.

Neway's sales volume would help us save the product. But a portion of the Neway business would eventually go to Goodyear, even as Firestone had acquired part of the GM bus business. But within a decade, General would drop out of the air spring business as U.S. Rubber (and, earlier, BF Goodrich) already had.

At Firestone, we attempted to build a market for air suspension at fleet level in our advertising and in personal visits to fleet executives. To explain the basics of air suspension technology during these missionary visits, and in the hope of piquing interest of these managers in the measurable monetary benefits that *Airide* offered to fleet operators, I produced a film, *Soft Touch*, with Tom Bank's considerable help. (I had been a filmmaker since high school and would elect early retirement twenty years later to found Cinemark, Inc.). Because we had little money to work with, I shot all of our footage with my 16mm Bolex, lighting factory scenes with a mixture of eight photofloods, a 750-watt Sun Gun, and daylight. Tom and I drew nine animation sequences on our own drawing boards to demonstrate graphically how air suspension systems worked.

One of the concepts we promoted was how air suspension is uniquely suited to provide *axle load equalization* on tandem-axle trailers to avoid exceeding axle load limits (18,000 lb [8182 kg] at the time, later raised to 21,000 lb [9545 kg] on interstates), and how it was possible to eliminate a thousand pounds of steel equalizing beams on tandem-axle trailers with a nine-foot spread.

In our marketing strategy we initially targeted two fleet markets: fragile cargo haulers and trucks or trailers that, by the nature of their business, traveled one way empty—both of which could realize major benefits from our technology. Fragile cargo haulers included government military electronics vans (which were loaded with fragile vacuum tubes at the time) and household moving companies—which frequently suffered damage claims due to constant over-the-road vibration on long-distance moves.

Typical one-way haulers were auto transport carriers (new car haulers), and—especially—tank trailers. Because the then-new, fragile, lightweight aluminum tankers could literally carry ten times as much cargo as the empty vehicle weighed, they had a history of structural damage when going home empty. Maintenance was particularly costly on such tankers because if a leak was discovered after a trailer loaded with gasoline had returned empty, the tank had to be thoroughly flushed to obviate any danger of explosion before it could be welded.

We showed our film to the management of E. Brook Matlack as well as Chemical Leaman Tank Lines, and our campaign met with some success. Fig. 8.13 is a 1963 photo of a Neway suspension on one of these tankers. Soon air springs began replacing leaf springs on most new tank trailers. By 1963, *Commercial Car Journal* somewhat grudgingly acknowledged that "Air suspensions, a bugaboo of the past, are making a gradual comeback. Most popular tractor applications so far appear to be on tanker fleets. It still offers the softest ride loaded or empty, and has been a big help in reducing tank damage caused by excessive pounding when returning empty."[6] Today it is rare to see a tanker on the highway that is *not* riding on air.

Partly as a result of our efforts, some household moving companies not only switched to air suspension, but both Mayflower and United Van Lines began displaying ads on the sides of their vans that read, "Soft Air Ride Suspension." Some members of Firestone's Law Department began fussing about this near-miss infringement of the Firestone "Airide" trademark. In a move a little like one of Art Hirtreiter's, I was able to persuade my management that we were getting free advertising for our product, and Vice President Russ Mitchell was successful in persuading the Law Department to stop harassing our customers' customers for bragging about their use of our product.

The variety of vehicle suspensions proliferated. At the top end in capacity was the 75,000 lb off-road suspension (Fig. 8.14) made by Ridewell Corporation of Springfield, Missouri. There were combinations of air springs and coil springs, air springs and rubber, and many trailing arm suspensions comprised of a leaf spring with an air spring at the end. American Carrier, a suspension manufacturer in Fresno, California, made it possible to convert existing leaf spring suspensions into this configuration. One of this company's more recent suspensions is pictured in Fig. 8.15.

Fig. 8.13 Neway suspension with Firestone air springs on a 1963 tank trailer.

Fig. 8.14 Firestone two-convolution air springs on an early 75,000 lb [34,000 kg] Ridewell suspension intended for off-road service.

Fig. 8.15 American Carrier's suspension kit makes it possible to convert existing leaf-spring suspensions to air.

In addition to those mentioned above, other early suspension manufacturers offering air springs in the late 1950s included Hutchins and Son, Western Unit, Turner QuickLift, Reyco, and Ridewell. Those using two-convolution air springs without a reservoir (and often with a negative lever arm) usually resulted in natural frequencies in the 135 to 145 cpm range.

Hutchens and Son of Springfield, Missouri, produced a unique *Hydro-Air* suspension for tandem axles which had a standard two-convolution air spring above each wheel filled with *liquid*, as seen in Fig. 8.16. The tops of these bellows opened into a closed structure that resembled the "air beam" on General Motors coaches—but it, too, was liquid-filled. As the suspension articulated, the fluid could readily flow through this large, horizontal conduit from one bellows to another. Midway between the two-convolution bellows, below the fluid-filled beam and opening into it, was a third (this one single-convolution) liquid-filled bellows, which sat on top of a second single-convolution part—this one filled with pressurized nitrogen. This single bellows constituted the spring in the suspension, flexing and further compressing the nitrogen as the working liquid dynamically deflected the fluid-filled bellows above it. The primary value of the system was articulation.

At Hendrickson, Hirtreiter's and Palguta's missionary efforts eventually succeeded in displacing the elongated General *Air Pillow* bellows and its expensive aluminum saddle reservoir with Goodyear's much more compact spring in a new Hendrickson offering.

After launching a joint venture with Turner QuickLift in the 1980s, Hendrickson Suspension Systems began manufacturing a trailing arm unit using **D**-type rolling lobe air springs. The company's current trailer suspension is pictured in Fig. 8.17.[7] Tailored for trucks, the Hendrickson suspension is widely used today, and is shown on a current Mack Truck in Figure 8.18.[8]

With the exception of special suspensions (e.g., the Watson & Chain *Hi-Lift* illustrated in Fig. 8.19,[9] or the similar Ridewell unit in Fig. 8.20) for which two and three-convolution bellows offer advantages, nearly all suspension manufacturers now use **D**-type rolling

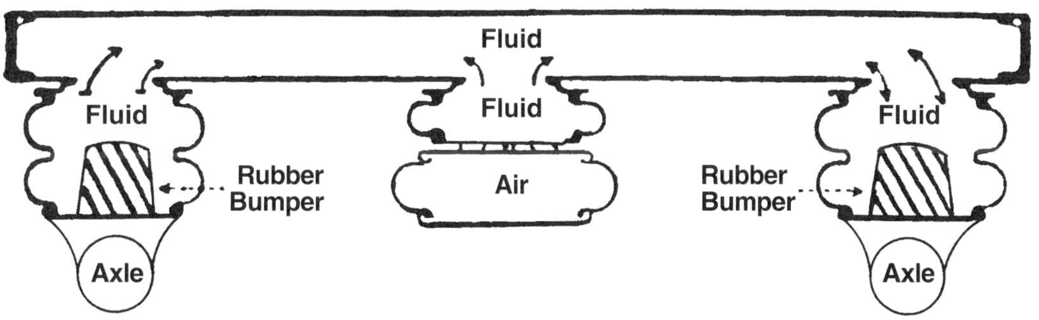

Fig. 8.16 A unique early hydropneumatic suspension for tandem axle vehicles by Hutchens and Son; its primary objective was improved articulation.

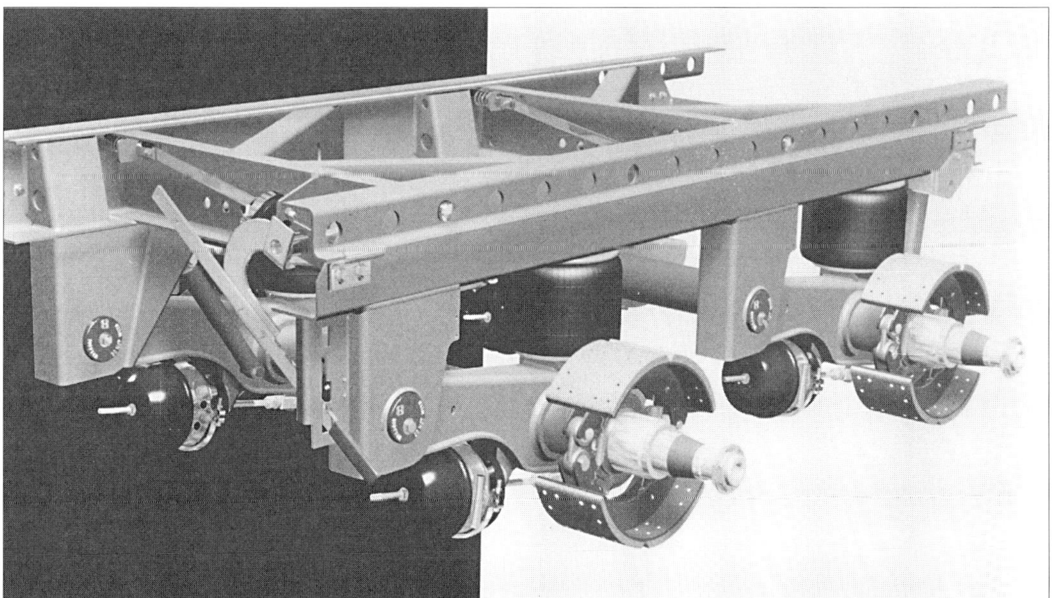

*Fig. 8.17 Modern Hendrickson trailing arm air suspension with Type **D** air springs made variously by Goodyear and by Firestone.*

Fig. 8.18 Hendrickson air suspension on a current Mack Truck.

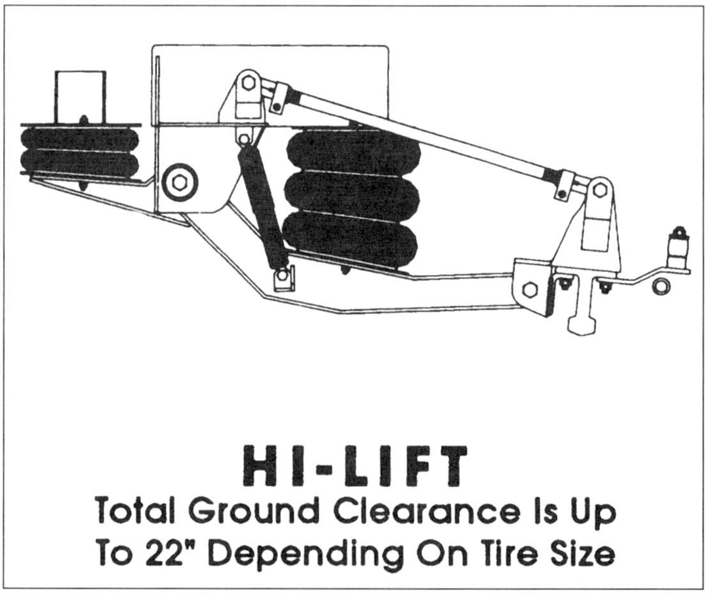

*Fig. 8.19 For some suspensions, e.g., this Watson & Chain Hi-Lift, two- and three-convolution bellows remain a better choice than Type **D** air springs.*

lobe (or "reversible sleeve") air springs—typified by the 1970 IHC trailing arm suspension depicted in Fig. 8.21,[10] and the early Ridewell unit pictured in Fig. 8.22. A current Peterbilt trailing arm truck suspension is shown in Fig. 8.23. Fig. 8.24 shows a Kenworth "eight-bag" KW 100 suspension. Fig. 8.25 shows a Freightliner suspension on another Kenworth. All of these suspensions have a natural frequency in the 90 cpm range.[11]

Fig. 8.20 Similar in function, this Ridewell unit also employs two- and three-convolution bellows to achieve a large excursion on its high-lift axle.

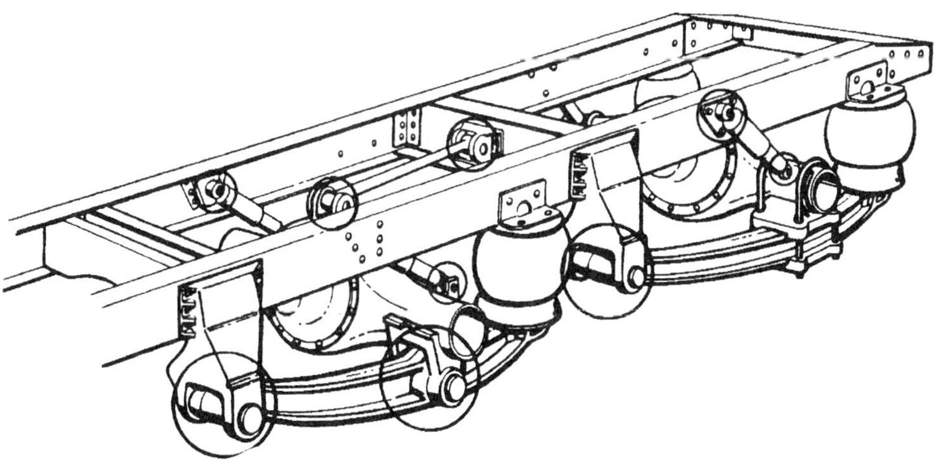

Fig. 8.21 The 1970 IHC trailing arm air suspension utilizing Type ***D*** *air springs.*

*Fig. 8.22 An early Ridewell unit with Type **D** air springs. Because of the very low hysteresis characteristic of air springs, all such suspensions employed shock absorbers.*

Fig. 8.23 Current Peterbilt trailing arm truck suspension.

Fig. 8.24 The Kenworth "eight bag" KW 100 suspension.

Fig. 8.25 Freightliner suspension on a Kenworth tractor.

Abandoning the tentative, cautionary view of air suspension taken by trade journals earlier, in his lengthy 1987 piece for *Owner Operator,* Kenneth Stadden described nearly all of the above suspensions, detailing the basics of the technology and its contribution to the operator's bottom line. In a later article after he became Technical Editor of *Heavy Duty Trucking*, he explained why suspending the truck's cab on air would easily pay out the added cost ($765 to $1030 at the time), telling readers where they could purchase the option, and pointing out that cab air suspension was standard equipment on trucks made by Ford, Freightliner, Mack, Peterbilt, Scania, Volvo, GM, and Western Star.[12]

By 1990, more than half of all Class 8 trucks and 20 percent of trailers on American highways rode on air suspension. As this is written, these figure have risen to 80 percent for trucks and 70 percent for trailers.[13]

Today most drivers sit on an air suspended seat in an air suspended cab on a truck chassis whose primary suspension is air. This proliferation of pneumatic springs on highway truck and trailer equipment is illustrated in Fig. 8.26. Type **D** reversible sleeve air springs

*Fig. 8.26 The proliferation of air springs on a modern tractor-trailer rig include Type **D** reversible sleeve air springs labeled **1**, the primary trailer suspension springs; the two-convolution bellows, **2**, are actuators that lift an axle on demand (with a control operated from the cab); air springs on the drive axle of the tractor are labeled **3**; **4** indicates the air springs on the tractor's front wheel suspension; smaller air springs, designated as **5**, are cab mounts; and **6** is the air spring under the driver's air suspended seat.*

labeled **1** are the primary trailer suspension springs. The two-convolution bellows, **2**, are actuators that lift an axle on demand, with a control that is operated from the cab. Air springs on the drive axle suspension of the tractor are labeled **3**. Springs labeled **4** are the springs for the front wheel suspension. The smaller air springs, designated as **5**, are cab mounts, and **6** is the air spring under the driver's seat.[14]

In a private survey of his own, Art Hirtreiter frequently asks long-distance truck drivers about their personal experience with riding on air (the above combination of chassis air suspension, pneumatic cab mounts, and air seats). On five separate occasions, these drivers have volunteered the opinion that not only do they feel better—"much more relaxed"—after a long drive, but they believe their working life as drivers has been extended by ten to fifteen years.[15]

The durability of the product has been remarkable—especially in comparison with steel springs. Gary Reynolds, Director of Sales and Marketing for Firestone Industrial Products, estimates that, after more than 40 years of product service over the years, replacement sales are only 5 to 8 percent.

During this same period of growth, after a decade-long hiatus since the advent of the X-trains, there was a resurgence of interest in air suspension for rail applications, led initially by high-speed train development in Europe and Japan. Spurred in this country by the U.S. Department of Transportation's subsidy support of mass transit systems (in an effort to reduce traffic congestion and air pollution from engine exhaust), new, sophisticated, electric rapid transit systems were built in many major American cities, including Chicago, Cleveland, Atlanta, Washington, Philadelphia, Baltimore, and even Dade County, Florida. All rode on air; some had air-coil systems. New, driverless automatic passenger transport systems between airport terminals, in theme parks, and on some university campuses were also built with air suspension.

By the 1980s, the trucks on new rail equipment for inter-city diesel-electric passenger trains, including Amtrak, were being built with air suspension. Typical is the rail truck shown in Fig. 8.27 with a pair of large, laterally-supported piston-type air springs, a design having elements of both Type **B** and Type **D**, situated midway between the axles. The Dofasco truck pictured in Fig. 8.28 went into service in Canada.

Rail air springs are the largest air springs of this configuration built to date (see Fig. 8.29). The spring on an Amtrak coach truck shown in Fig. 8.30 is 640 mm (25.2 in.) in diameter and has a load range of 70–145 kN (16,000–32,600 lb). Its natural frequency, depending

*Fig. 8.27 A 1980s rail truck with a Firestone air spring midway between the trucks. These air springs, having design elements common to both Type **B** and Type **D**, are externally supported circumferentially to increase lateral as well as fore-and-aft stability.*

Fig. 8.28 This Dofasco rail truck in service on the Canadian National Railway employs a unit of similar design.

Fig. 8.29 Air springs designed for rail applications are the largest that Firestone has manufactured to date.

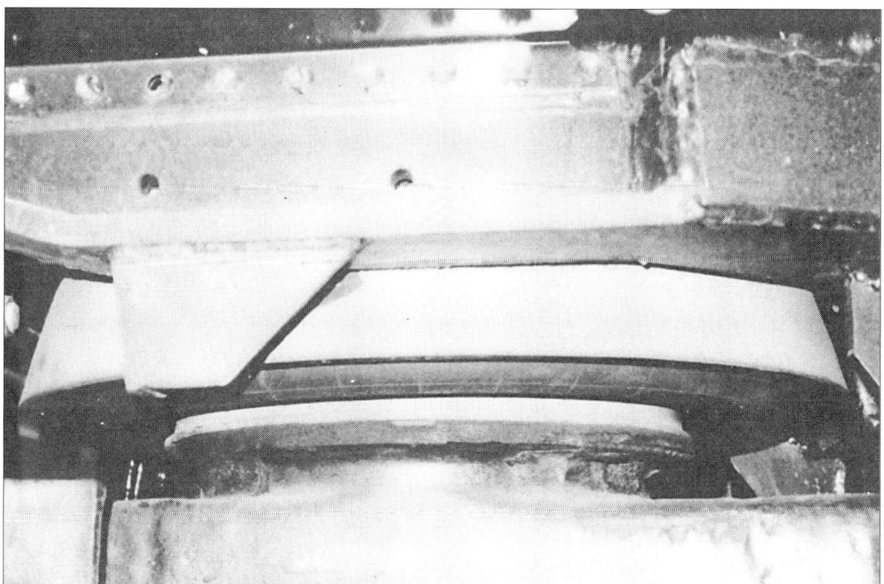

Fig. 8.30 This Amtrak air spring, 640 mm [25.2 in.] in diameter, has a load range of 70–145 kN [16,000–32,600 lb]. Its natural frequency, depending on reservoir volume, ranges from 45 to 70 cpm.

Fig. 8.31 Components of a typical rail air spring assembly include an aluminum piston, the reinforced-rubber air spring, a bead ring, bead skirt (which controls the lateral rate of the spring), an internal bumper stop, and a mounting plate.

on reservoir volume, has a range of 45 to 70 cpm (45 cpm with a reservoir volume of 49,200 cm^3 and 70 cpm with none). Components of a typical rail air spring assembly (Fig. 8.31) include, clockwise from upper left, the aluminum piston, the reinforced-rubber air spring, bead ring, bead skirt (which controls the lateral rate of the spring), the internal bumper stop, and the mounting plate. The sales volume of air springs for rail applications (including Great Britain and Continental Europe as well as the United States—and more recently Australia and Brazil), has become prolific enough that Firestone has added *Airail*™ to its product line.

Other diverse automotive suspension applications emerged over the years. One of the more interesting was the General Motors motor home, made by GM's Truck & Coach Division in Pontiac—a recreational vehicle that debuted in 1973. Sporting a Toronado front wheel drive with coil springs, the unique rear suspension of the

*Fig. 8.32 A unique double-ended Type **D** "banana" air spring was manufactured by Firestone for the tandem-axle General Motors motor home.*

tandem-axle vehicle is shown in Fig. 8.32. Its geometry was similar to that of the 1960s Trailmobile air suspension, but the horizontal two-convolution air spring used on those highway trailers would have resulted in an unacceptable ride for a passenger vehicle. Instead, the motor home had double, opposed piston "banana" air spring shown—a Firestone development.

For upscale (i.e., motor home) campers who had spent early evenings in the gathering dusk shimming their truck's wheels with rocks and fallen tree branches in an effort to level up their Servel-type gas refrigerators, the GM motor home had a feature that seemed heaven sent. After parking for the night, the driver had merely to turn his attention to the small lever on the console to his right—manipulating the little joystick until the bubble was in the center of the circle on the adjacent leveling gauge.

The GM motor home met its demise in 1978, a casualty of a national recession in combination with the "Arab oil boycott," which resulted in a national gasoline crisis and long, frustrating lines at most filling stations. Gas mileage (e.g., 8–9 mpg) had not been a selling point for motor homes.

We began our campaign to attract new business for air springs with an evangelistic fervor. Both Firestone and Goodyear vigorously pursued non-automotive applications for our product. At Firestone, we began featuring a line of pneumatic springs for *isolation* applications called *Airmounts*®—many produced in the same molds as *Airide* springs for vehicle suspensions. Before I arrived on the scene from Detroit, field engineers Mike Kray, Jerry Marsh, and Bob Weir had already developed a substantial base of "trade sales" *Airmount* business. The modus operandi of these engineers (like that of Art Hirtreiter, Tom Burkley, and Andy Palguta at Goodyear) had been to help customer engineers solve their problems using air spring technology.

Because air springs made it possible to package a large effective static deflection in a remarkably small package (see Chapter 4), producing lower natural frequencies than could be practically achieved by any other kind of vibration isolation device, one common non-automotive application became *machinery isolation*—either means to keep a machine from disturbing its surroundings, or finding a way to isolate sensitive apparatus from a disturbing or damaging environment. Fig. 8.33, for example, shows a Bethlehem Tool Company 25,000 lb forming press whose impacts had shaken

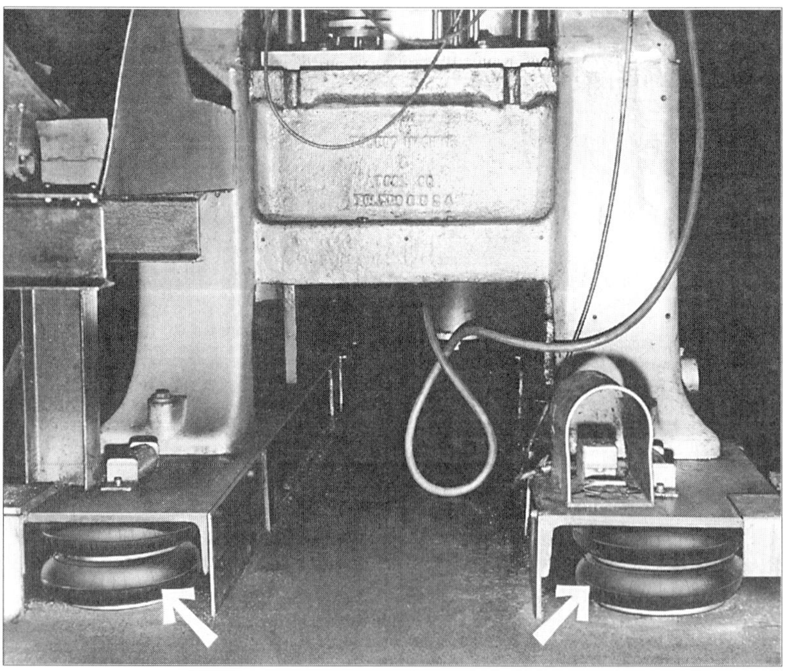

Fig. 8.33 Two-convolution air springs beneath this 25,000 lb [11,400 kg] Bethlehem Tool forming press reduced seismographic recordings of its impact shock into factory buildings by 90 percent.

factory buildings at intolerable levels until they were mounted on two-convolution air springs—which typically reduced seismographic recordings in factory building by more than 90 percent.

Mounting vibrating feed conveyors and shaker screens (e.g., those manufactured by Link Belt, Carrier Conveyor, and Allis Chalmers) on air springs speeded up throughput while performing their isolation function. The iron ore screen in Fig. 8.34, manufactured by Canadian Allis Chalmers, processes 1500 to 2000 tons per hour.

The largest Firestone *Airmount* made to date is the 37 inch (94 cm) diameter air spring pictured in Fig. 8.35. A better feeling for its size may be had from Fig. 8.36. These springs isolate a 100,000 lb (45,000 kg) concrete slab that functions as inertial ballast for seismographic equipment in earthquake-prone Japan. The three-convolution air spring has a natural frequency in the 80–90 cpm range. It is anchored and sealed with bead rings and (40) bolts instead of crimped bead plates because a crimping die of this size has yet to be made.

Fig. 8.34 Another isolation application was this Canadian Allis Chalmers vibrating feed conveyor, which processed 1500 to 2000 lb [680 to 910 kg] of iron ore per hour.

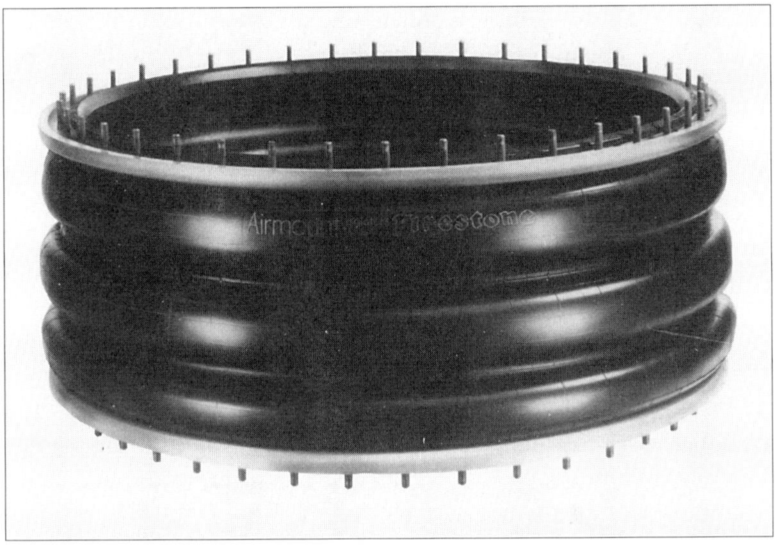

Fig. 8.35 Largest multi-convolution air spring manufactured by Firestone to date is this 37 inch (94 cm) diameter Airmount developed to support a 100,000 lb (45,000 kg) concrete slab on earthquake-sensing seismographic equipment in Japan.

Fig. 8.36 Demonstrating the size of the Airmount developed for a Japanese seismograph.

At the other end of the scale is the 6 inch [15 cm] single-convolution spring pictured in Fig. 8.37. In 1958, Mike Kray had helped Western Electric grow perfect germanium crystals at the beginning of the transistor age using a similar spring having a natural frequency of about 185 cpm (Fig. 8.38). With the subsequent development of 1-X (Type **B**) springs, a 5.2 inch [12 cm] diameter air spring with a much lower natural frequency (80 cpm) is now available for such applications (Fig. 8.39).

Fig. 8.37 Smallest bellows-type is this 6 inch [15 cm] single convolution spring with crimped bead plates and blind nuts.

Fig. 8.38 At the beginning of the transistor age in the 1950s, single-convolution Airmounts isolated this Western Electric germanium crystal growing apparatus from ambient vibration.

Our group worked on a wide variety of isolation applications. We floated one entire floor of Houston's then-new Humble Oil Building on air—isolating all of the air conditioning equipment and other machinery from the structure of the building. At the University of Southern California, I helped an architect install a battery of concrete acoustical chambers, achieving 99.9 percent isolation in the sonic range. The following year, when officials of the Seattle World's Fair (justifiably) complained about the ride on the Alweg Monorail—which had a suspension that used two-convolution (German-made) air springs with no reservoir on a suspension with a negative lever arm, we replaced them with Type **D** reversible sleeve air springs, dropping the natural frequency by more than 40 cpm.

We worked on some of these applications primarily for the publicity—so we could demonstrate what could be done with air springs. The applications, and the problems we managed to help solve, made spectacular ads in *Design News* and *Product Engineering*—and sometimes these reached engineers who were completely unaware of the product. There were companies at the time who refused to let any technical representatives call on their engineering departments without the approval of the purchasing department. Each of the engineers who responded to one of our ads got a personal letter (in an age before word processors) and a copy of the air spring Design Manual that Tom Bank and I put together and published in-house on our division's Multigraph printer.

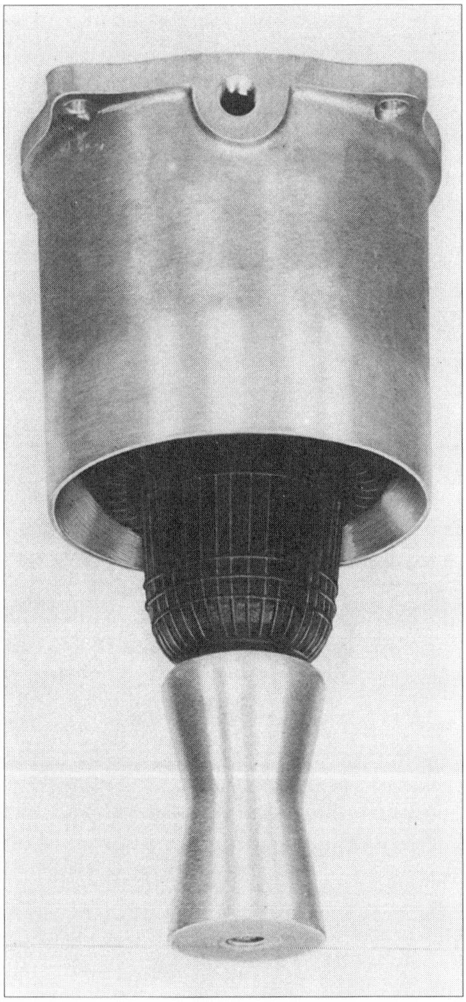

*Fig. 8.39 A smaller (externally-supported) Type **B** air spring for such applications today provides a much lower natural frequency (80 cpm vs. 158).*

At Goodyear, Andy Palguta, Art Hirtreiter and his staff were solving similar problems—one of which involved a storage and handling container for the Polaris Missile, which obviously had to be protected from bumps in transit. Making air springs in the form of pressurized, interconnected "logs," eight of which were radially disposed around the missile inside the huge container (Fig. 8.40), Goodyear's accelerometers demonstrated that the system absorbed 85 percent of impact shock if the container should be dropped.

In a related program, the Goodyear group designed rolling lobe springs for the suspensions of Atlas Missile Transporters (Fig. 8.41).

Fig. 8.40 Surrounded by Goodyear pneumatic "logs," Polaris Missiles were protected from bumps in transit.

They later isolated NASA's Space Shuttle (Fig. 8.42) so a series of sensitive tests could be run on its instrumentation without interference from environmental noise. (Dwarfed by the Space Shuttle, the rolling lobe springs are barely visible in lower right of the picture; see close-up in Fig. 8.43). Perhaps the most spectacular feat of the Goodyear team was their fabricating a single rolling lobe air spring 32 feet [9.8 m] in diameter to support NASA's 363 ft [110 m] Saturn rocket, weighing 6.2 million lb [2800 metric tons] on its launch pad.

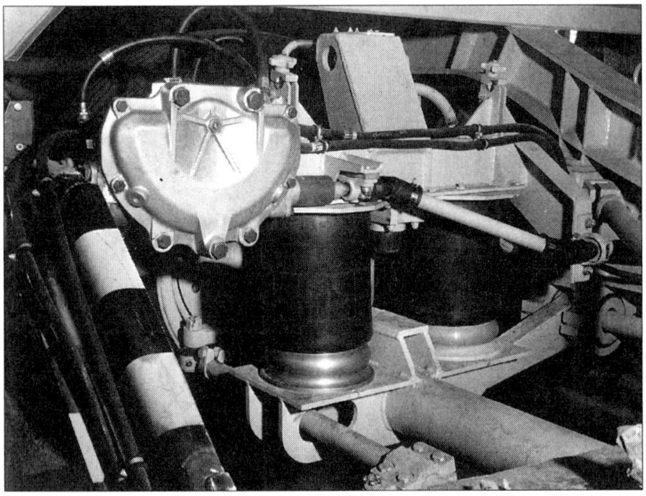

Fig. 8.41 Goodyear rolling lobe air springs on the suspension of the Atlas Missile Transporter.

Fig. 8.42 Goodyear rolling lobes also isolated NASA's Space Shuttle from its environment so that sensitive tests could be run on the Shuttle's instrumentation free of ambient vibration.

*Fig. 8.43 Detail of the Type **D** air springs isolating the Space Shuttle.*

Several years later, Art Hirtreiter was called upon to solve a related problem for the U.S. Navy. To measure vibration in the turbines of jet fighter planes, it had been standard operating procedure to periodically transport the aircraft to shore for such tests—putting these planes out of service. At the Navy's request, Art designed the massive jet engine test platform pictured in Fig. 8.44, making it possible to conduct these tests aboard the aircraft carrier.

The jet engine test facility was supported on the deck of the ship by ten rolling lobe air springs, isolating it from ship vibrations while also maintaining a constant platform height and lateral position with reference to the deck of the ship, regardless of the engine weight or thrust of the engines—or the roll and pitching movements of the ship. Additional air springs, horizontally mounted, provided lateral stability at the corners; and a bank of eight more units absorbed the thrust and the axial vibration of the engine under test. The system had a complex pneumatic system with four vertical sensing valves, two thrust sensing valves, an air pressure regulator and gauge, and seven air check valves. Hirtreiter's group also wrote the operating manual for the facility.

In contrast to these *isolation* applications for air springs, all of which utilized natural frequencies well below those of the disturbing frequencies to be isolated, Carrier Conveyor's ingenious chief engineer, Albert Musschoot, began exploiting the effects of forced-frequency to natural-frequency ratios near *unity* (see the resulting peak in Fig. 4.1, Chapter 4) to *amplify* vibration. Musschoot patented a scheme for *tuning* the amount of vibration amplification on a conveyor by using *pairs* of air springs in opposition as a compound spring. By adjusting the pressure in the springs, the amplitude of the conveyors made by the Louisville company (and consequently the rate of flow), could be adjusted.

Earlier, at Link Belt, Musschoot had designed and overseen the installation of eight *Kar-Flo* freight car unloading facilities (shown schematically in Fig. 8.45) which worked on the same principle. Rocking on 60 foot [18 m] steel beams, half-million-pound [quarter million kg] grain cars were emptied in eight minutes, the pressure in the air springs automatically changing as the load became lighter. The motion shook the grain to the center of the car and out the door, cutting unloading time by more than 70 percent.

Musschoot later founded General Kinematics in Barrington, Illinois, a company that makes giant coal conveyors that operate on controlled resonance. The same principle, Musschoot has pointed out, could be used to operate pile drivers, or even jack hammers (Fig. 8.46).

In an adventure similar to Art Hirtreiter's, I became involved in a military project with the Strategic Air Command (SAC) in the early 1960s. It was the height of the Cold War. Some families in Indiana were actually digging fallout shelters in their backyards and stocking them with food. But my experience had a comic twist.

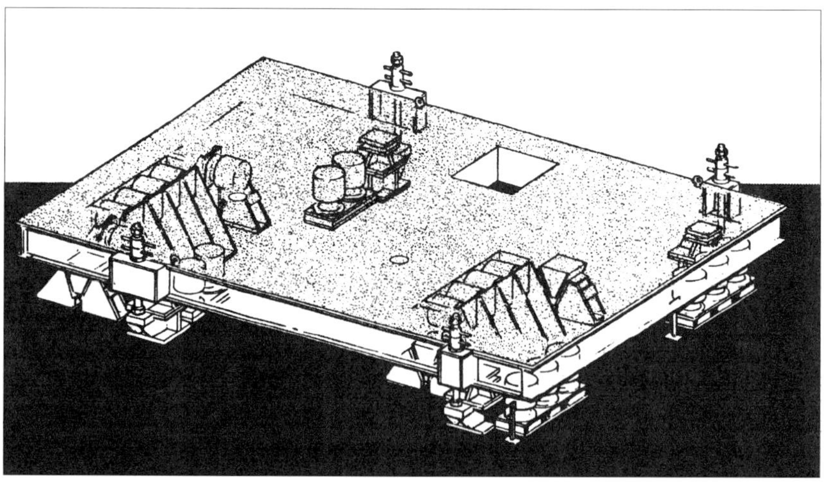

Fig. 8.44 Art Hirtreiter's jet engine test platform which used a battery of automatically-adjusted rolling lobe air springs to isolate engines of carrier aircraft from vibration, and even from the roll of the ship on which tests took place, making it unnecessary for carrier jets to be periodically returned to the mainland for engine evaluation.

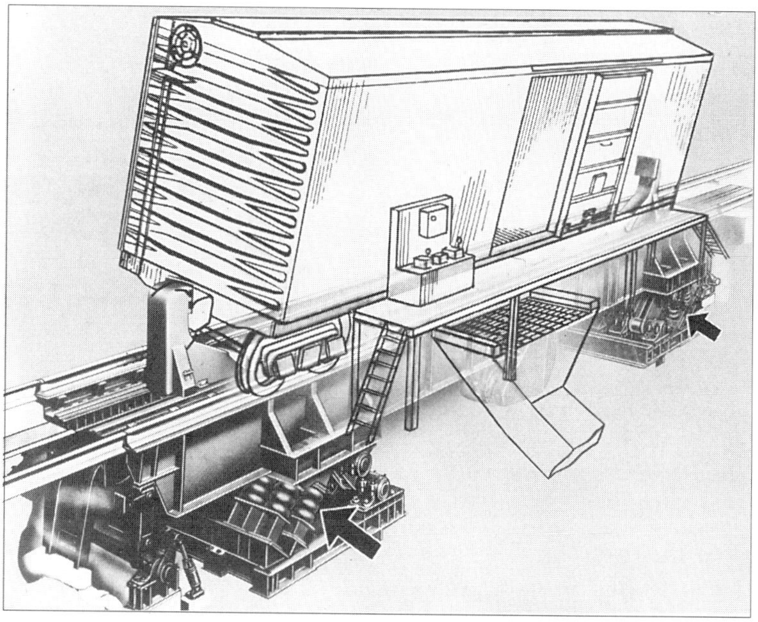

Fig. 8.45 Using Firestone air springs to **amplify** vibration, Link Belt's **Kar-Flo** conveyor shook grain to the center of a freight car and out the door, cutting unloading time by more than 70 percent.

Fig. 8.46 A jack hammer application for two-convolution air springs.

To sweep the skies with radar while keeping their scanning sites mobile, SAC operated fleets of trailers filled with electronic equipment, some with rotating radar antennas on their roofs. The trailers moved as a truck convoy whenever a new location was ordered. But the stiff leaf spring suspensions on the SAC vehicles were less than kind to the 20,000 vacuum tubes inside the two dozen trailers (again, this was before the advent of transistors). Each time the vehicles were moved to a new location, they invariably knocked out some of the fragile tubes, and there followed the time-consuming task of manually plugging thousands of them into test equipment, taking as long as a week to find all of the defective ones—putting the radar out of action in the meantime. What if the Russians were coming?

SAC said they wanted to make a "radar *train*," removing the existing suspensions from the trailers and mounting them on flat-bed rail cars (Fig. 8.47). They asked whether we could "float them on something."

I flew to Ogden, Utah, where the train was being built in a government rail depot. In addition to providing needed isolation, our air springs would, I explained, give them the advantage of being able to deflate the system upon arrival at a new site, providing a firm mounting for the PPI scanning antennas. I went back to my motel room with my design manual, did some calculations, and later that night made some sketches.

But I felt we needed to know the natural frequency of the loaded flat car on its coil spring trucks to be sure the natural frequency of my air spring suspension was far enough below this value that we wouldn't have any resonant interaction. But the Ogden personnel had no data on the rates of the rail car springs. They did indicate, however, that a work-car coach on the site had the same trucks, and it weighed about as much as each of the flat cars would after loading.

But how to jostle one end of a whole railroad car to determine its natural frequency? In one of his books, Physicist Georg Gamow described a thought experiment in which the entire population of China jumped in unison to see if they could perturb the orbit of the Earth.

I experienced an inventor's "Aha!" To empirically check the suspension frequency of the coach at the rail depot, I was actually permitted to perform Gamow's experiment—albeit with a cast of forty instead of a billion.

Emptying out the shops at the rail depot, my hosts graciously provided a wide assortment (some of them *quite* wide) of mostly middle-age males, many clad in the bib overalls common in rail shops, who jumped up and down in unison on command—leaping to my accelerating hand-claps until we had determined that the natural frequency of the trucks was much higher than that of my proposed isolation system.

Fig. 8.47 Goal of the Strategic Air Command during the Cold War of the early 1960s was to take radar trailers loaded with fragile vacuum tubes off the highway and somehow isolate them on a radar ***train*** *to avoid several days of radar downtime while searching for damaged radio tubes after each relocation.*

Fig. 8.48 The author's isolation system for SAC's radar train, which was built in Ogden Utah. Each of the electronics trailers was supported at the corners on two-convolution air springs, with horizontally-mounted three-convolution bellows fore and aft to absorb the shock of coupling. Other two-convolution springs provided lateral stability.

It was quite a spectacle, that sea of well-fed Mormon stomachs, each bouncing at its own natural frequency. I could not have guessed then that, fifteen years later, Albert W. Foster would actually quantify such anatomical values while developing cab mountings for Ford trucks (see Chapter 7).

As the installation was built, each of the electronics trailers was supported at the corners on two-convolution air springs, with horizontally-mounted three-convolution bellows fore and aft to absorb the shock of collision when coupling up the train, and pairs of two-convolutions for lateral stability. The result is pictured in Fig. 8.48. After the train had completed several moves, we were pleased to receive SAC's report that tube failures during transit had disappeared.

The properties of flexible pneumatic springs and their molded configurations have spawned unanticipated *actuator* applications, typified by the press in Fig. 8.49, the first of which was built by Tom Bank—one that has since been duplicated in many customer factories. The large two-convolution bellows are used to close the press and apply pressure, while smaller ones open the platens.

In the Design Manual we published, Tom and his draftsmen prepared a set of drawings illustrating how air springs could be used in actuator applications. These pages from the manual are reproduced in Fig. 8.50. Fig. 8.51 shows a small single-convolution air spring functioning as an actuator for a locking mechanism on a machine that cycles several times a minute.

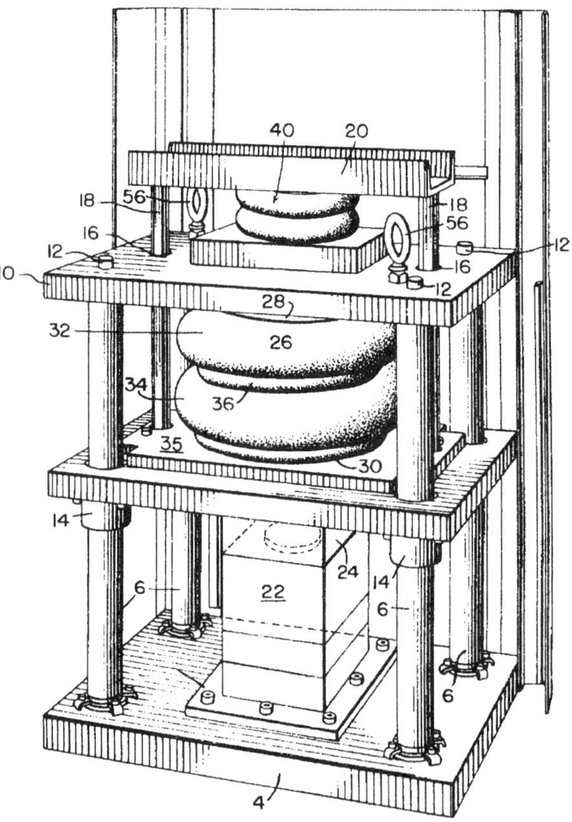

Fig. 8.49 Tom Bank's design for a pneumatic press, using large two-convolution air springs for the requisite pressure to close the press and smaller ones to open it.

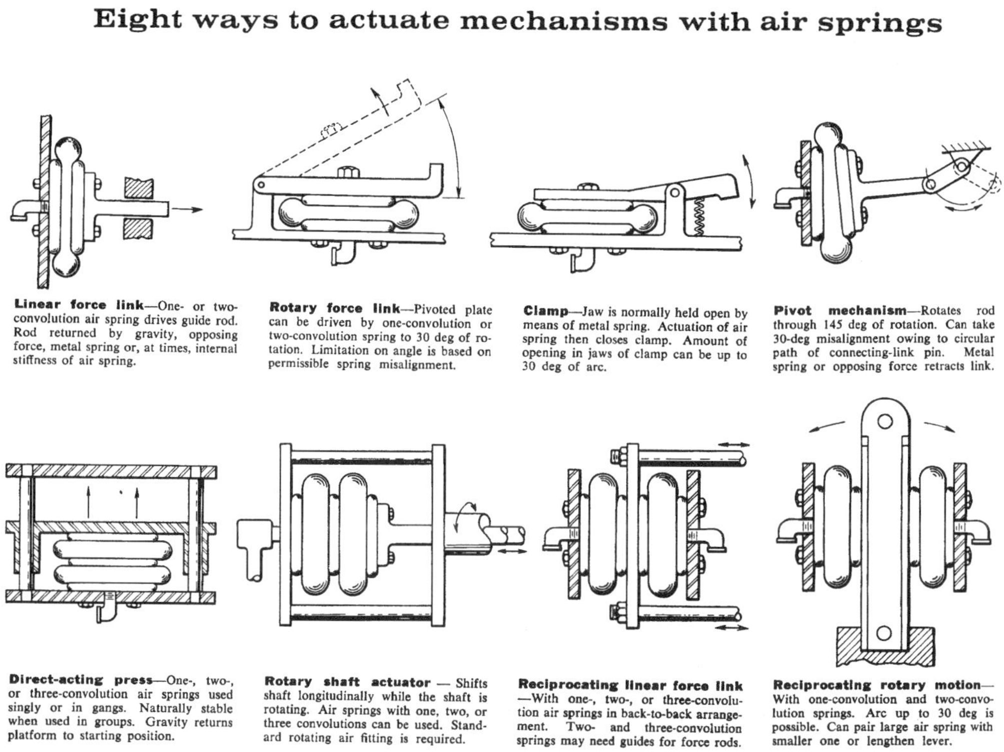

Fig. 8.50 Linkages for actuator applications of air springs, as illustrated in an early Firestone Design Manual.

Fig. 8.51 A simple actuator application using a small single-convolution air spring operated an automatic locking mechanism on a factory machine.

Unlike air cylinders with sliding cup seals, air spring actuators introduce *no friction*. As early as 1960, this characteristic began to make them popular in the paper industry. Pioneered by Beloit Corporation, the giant stone rolls on paper calenders are floated on *Airmounts* (Fig. 8.52), which also precisely control rolling pressure as they squeeze water from the slurry pulp. A stack of pressure-controlling two-convolution air springs on a 206 inch [4.24 m] newsprint calender is shown in Fig. 8.53.

Fig. 8.52 These heavy stone rolls on a Beloit paper calender float on Firestone Airmounts, which also control the amount of roll pressure on the machine.

Fig. 8.53 Stack of pressure-controlling two-convolution actuators on a 206 inch [4.24 m] newsprint calender.

The smallest *linear* actuator to date is shown in Fig. 8.54. Molded in a configuration similar to the original 2-X Hollis design (the ends buckle inward to form two opposing pistons), the compact unit has a one inch [25 mm] stroke. But Firestone has since developed *circumferential* actuators. An *Airgripper*™ and an *Airpicker*® are illustrated in Fig. 8.55. When inflated, the inside diameter of the hollow *Airgripper* abruptly decreases, squeezing around any rod-like projection (e.g., the electron gun sticking up from the back of a TV picture tube), permitting automatic machinery to lift and safely transport the object. Performing the reverse function, the pencil-size *Airpicker* inflates, swelling up inside the neck of a bottle, for example, to lift it off an assembly line.

Fig. 8.54 This smallest Firestone linear actuator has a one-inch [25 cm] stroke (the ends collapse inward as in an early 2-X design).

The proliferation of air spring designs and their multiplicity of applications over the years has resulted in the wide array of sizes and configurations, sampled in Fig. 8.56.

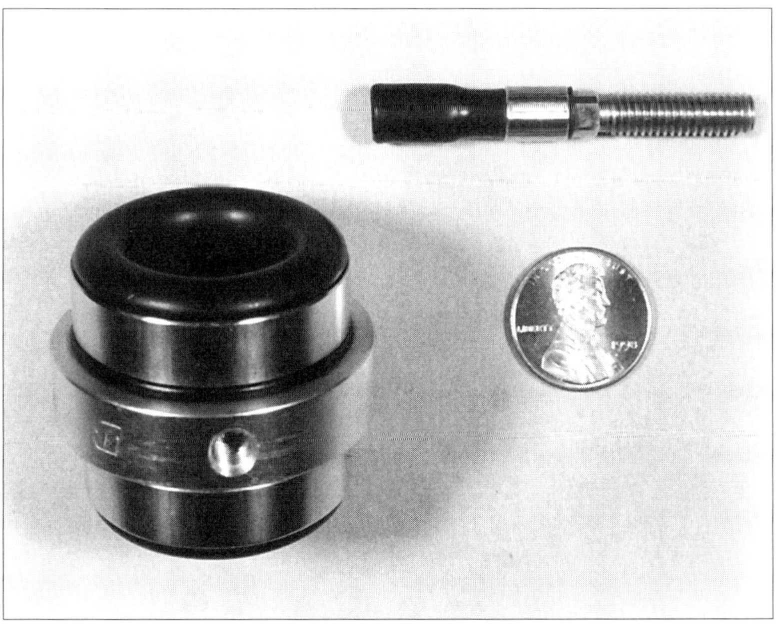

*Fig. 8.55 Firestone **Airpicker** and **Airgripper** circumferential actuators inflate to automatically pick up objects from assembly lines.*

Fig. 8.56 Showing the wide variety of air spring sizes and designs that have been developed since air suspension became a commercial reality half a century ago.

To state what should, by now, be obvious (although it wasn't that obvious at the time!), we did manage to attain our first million dollars in sales at Firestone in that fateful 1960 fiscal year. Goodyear was also doing well, and General Tire was making solid inroads with its wide girdle band Type **C** spring. By the third year, our sales had doubled. They would double again by 1968. Although General has since dropped out, a number of European manufacturers have added air springs to their product line. Some (Continental, Phönix Gummiwerke, Pneubras, Pirelli, Dunlop) are doing well.

Air suspension has continued to grow every year since 1960, and today, coming up on the golden anniversary of their first successful commercial application on GM buses, the worldwide market for air springs, exclusive of all attendant hardware, is approaching half a billion dollars.

Chapter Nine

THE FRENCH CONNECTION

In contrast to the United States, where commercial air suspension began on buses, to be followed by truck and trailer applications, in Europe pneumatic and *hydropneumatic* springing began with passenger cars. Indeed, it began as early as 1905 as an aftermarket device in France: the *Aeros* "Balloon" was an inflatable rubber ball that was placed between the semi-elliptic leaf spring and the frame of a car. French engineer George Messier later developed a hydropneumatic spring for the passenger car that bore his name from 1922 to 1930. Messier's oleomatic shock absorber, in fact, played a role in the early development of the Citroën suspension, which began in 1939. Over the years, Mercedes, BMW, Renault, Borgward, and the Land Rover have all had flings with air suspension.[1]

In Paris, in 1953, the same year General Motors introduced its first parlor coach on air suspension—five years before air spring technology emerged on American passenger cars—Citroën announced its 15 cv with hydropneumatic suspension on the rear wheels. The following year Citroën brought out "the car of the new era," the DS 19, with hydraulically-assisted steering, hydraulically-assisted brakes, and with hydropneumatic suspension front and rear—all powered by the same engine-driven hydraulic pump.

A hydropneumatic suspension is a combination of an oil shock absorber and a sealed pneumatic spring. Its working fluid is the shock absorber oil, which communicates wheel motion to the pneumatic spring element by working back and forth against the pressure of a compressed gas in a sealed chamber. The gas is the spring element in the system.

With reference to Fig. 9.1, a cross section of the front suspension of the Citroën, the round metal accumulator, or "suspension sphere" (**4**), at the top of the suspension unit is charged at the factory with a fixed amount of nitrogen under very high pressure (because the effective area of the spring is limited to the piston area of the shock absorber). In operation, vertical wheel motion is transferred by the suspension arm (**7**) to the piston in the shock absorber cylinder (**5**), causing shock absorber oil to pass back and forth into and out of the accumulator. Any movement of the suspension thus further compresses or expands the gas—from which the oil is separated by a flexible diaphragm (represented by the heavy, wavy black line) inside the sphere. Nitrogen is used instead of air in order to exclude any potential corrosion problems that might be associated with the use of oxygen.[2]

Operating pressures of the Citroën passenger car system were much higher than those of conventional pneumatic suspensions—in the 2500 psi [175 bar] range. But this pressure constituted no problem for the flexible rubber element in the accumulator, because the

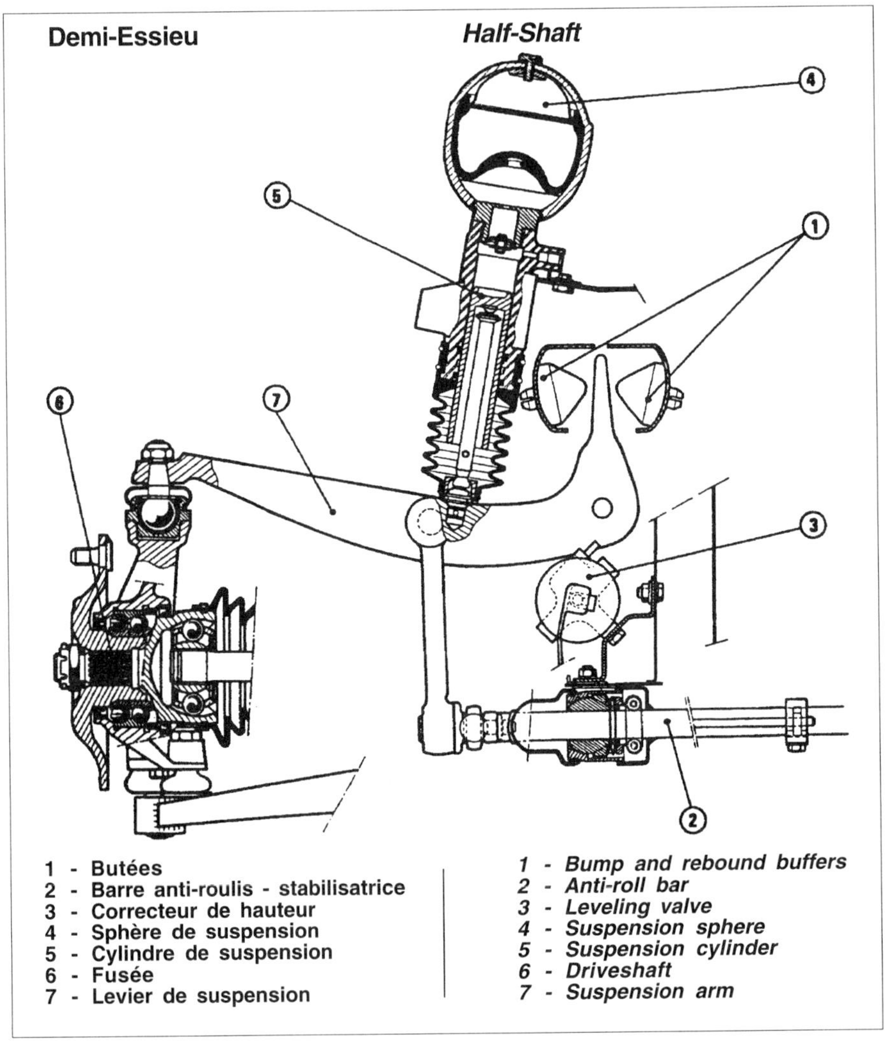

Fig. 9.1 Cross section of the front hydropneumatic suspension of the 1954 Citroën.

pressure on both sides of the diaphragm separating the two fluids was the same. Early Citroëns had two leveling valves, or "height correctors" (**3**), one each, front and rear. These contained a sliding valve that, when load was added to the vehicle (further compressing the gas, thus decreasing the ground clearance), opened the high-pressure hydraulic line from the main pressure reserve, bleeding in hydraulic fluid until the ride height of the vehicle was restored. Similarly, when people got out of the car, the valve discharged oil to the storage tank. Like the Delco leveling valves on GMC buses, these height corrector valves contained a dashpot arrangement which delayed the actual opening of the valve during normal jouncing movements of the suspension.

As an added convenience, the Citroën had a manual control available to the driver which permitted him to adjust the static height of the car. The action of the device, as described in the vintage Citroën operator's manual begins, "The control lever on the console acts on a linkage that simultaneously moves the slide valves of the front and rear height control correctors."[2] The driver's manual suggests that this option be used to make tire changing easier or to optimize the aerodynamics of the car. The feature would be offered 40 years later on the air-suspended Lincoln Mark VIII, but the Lincoln system would be automatic; it would lower the ground clearance of the car at highway speeds.

The Citroën had spectacular ride and handling characteristics, and it might have been more popular in the United States had it been less expensive and had its styling not been so radically odd by American standards of the 1950s and '60s. Since Citroën had interconnected the front and rear steel springs on its 1950 2 cv model, interposing another (horizontal) coil spring in between to control the pitch on this short-wheelbase car, it is surprising that the company elected not to interconnect its front and rear hydropneumatic suspensions to reduce brake dive and acceleration squat on its initial DS 19—nor did it hydraulically inhibit lateral roll, relying instead on a heavy roll bar.[3]

Engineering icon Maurice Olley of Rolls Royce, and later of General Motors Technical Staff (and perhaps the first engineer to analyze in detail the interdependence of front and rear suspension characteristics) was a fan of hydropneumatic suspensions for this very reason: because they offered the possibility of interlinking the front and rear springs of a vehicle to control pitching movements. (Like the earlier Citroën, the American Packard Clipper Custom of the mid 1950s had also accomplished such a function mechanically, using interconnected *torsion bars*—a system engineered by William Allison).

As British engineer Joseph Lowrey explained in a 1956 article in *The Motor*, "The primary object of interconnected front and rear suspensions is to give the car a flat ride, without any pitching action. In essence, when first the front wheels and then the rear wheels of an ordinary car pass over a bump, there is inevitably first a tendency for the front of the car to rise and then for the rear to do likewise. But, if a car could be supported like a child's see-saw, at a single point under the center of gravity, each bump would lift it bodily twice, but would not produce any pitching tendency."[4]

As recounted in Chapter 5, Chrysler's Director of Advanced Engineering, Otto Winkelman (who, before World War II, was with Mercedes Benz Racing Division) wanted to employ air suspension for this purpose in the 1950s when Art Hirtreiter was introducing his rolling lobe air spring to Chrysler.

"The front-to-rear connecting medium was to be air, with a complicated control valve, together with height controls," Hirtreiter recalls. "The main weakness of the system was that [compared to hydraulic systems] air moves too slowly, and the signal from the front to the rear of the car was too late. I tried relay valves such as are used on the rear brakes of 18-wheelers, but the system was not successful—or practical—and the project was dropped."

Fig. 9.2 illustrates in principle how this anti-pitching function can be accomplished with a hydropneumatic suspension. British writer L. J. K. Setright chronicles European developments in this regard in his 1977 book, *Anatomy of an Automobile*:

"British Leyland have adopted two ... complex systems of this type, both designed by Dr. Alex Moulton. The first, the Hydrolastic system, appeared in 1962 on the Austin 1100, and was subsequently adopted for a number of other models [including the Morris Minor, Cooper, and Austin Mini]. In the Hydrolastic arrangement, each wheel has its own suspension unit comprising a rubber springing element and a fluid chamber with integral damping valves; the interconnection between the front and rear chambers on each side is by means of a fluid-filled pipe. The upward movement of one wheel displaces fluid through one damping valve and along the pipe to the other unit, which is thereby extended to compensate for the compression of the first one.

"This principle has been refined in the more recent Hydragas system introduced in 1973 on the Austin Allegro and later fitted to the Corporation's 18/22 cars. The rubber side of the suspension units has been replaced by gas under compression, to enhance the spring response, and the damping and other details have been improved also. The effect is to make possible independent provision for soft ride and good handling stability."[5]

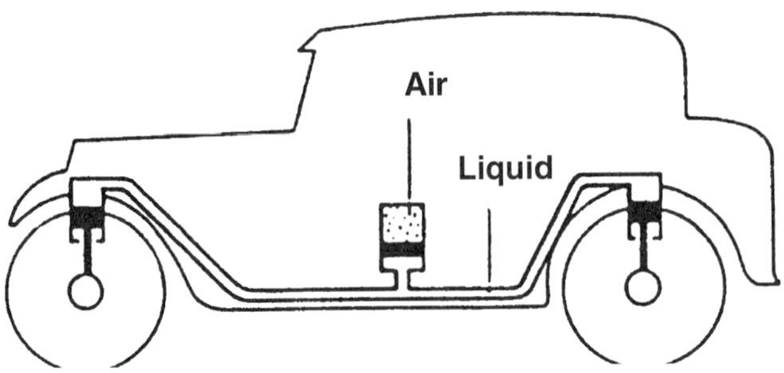

Fig. 9.2 Showing in principle how an anti-pitching function can be achieved with a hydropneumatic suspension.

(To which, Setright might have added, r*egardless of ambient temperatures.* As Art Hirtreiter points out, "in contrast to Hydragas systems, which are not temperature sensitive, American owners of Mini Minors, Cooper Minis, Austin Minis, and similar cars having Hydrolastic suspensions—that became rock hard in cold weather—usually put their cars in storage for the winter because their ride became intolerable." Not only did the oil become more viscous, but the rubber "biscuits" whose distortion displaced the oil, became impossibly stiff in American winters.)

"Citroën and the Hydragas systems, and a related system on the Mercedes-Benz 450 SEL 6.9," Setright continues, "rely on a constant *mass* of compressed gas—actually nitrogen—acting as the spring medium. Another form of pneumatic suspension uses a constant *volume* of gas. This gives a different character to the ride: constant mass gives a rising spring rate (it gets stiffer as it is deflected) allowing soft springing in normal conditions without excessive wheel travel under heavy loads, whereas the other type gives a virtually constant rate."

Art Hirtreiter explains why this is so: "One of the controlling factors affecting the rate or the natural vibration frequency of pneumatic springs is the *actual volume* of the elastic medium, whether it be gas or air. As the load on the suspension system is increased, the diaphragm of the constant *mass* spring (e.g., the Citroën) compresses the gas into a smaller volume, and the rate as well as the natural frequency of the system increase proportionately. For the constant *volume* spring, as the load is increased, additional gas (or air) is pumped into the spring at a higher pressure. The height, and therefore the volume, the spring rate, and the natural frequency remain constant while the load-carrying ability is increased in proportion to the pressure. With the constant volume type pneumatic spring, contouring the piston for a 1-X reversible diaphragm [Type **B**] air spring or a rolling lobe [Type **D**] part can also provide a rapid change in effective area and, consequently, an increase in spring rate with deflection.

"Back in the '50s," Hirtreiter adds, "I discussed with Maurice Olley the possibilities of a constant frequency suspension consisting of a constant mass gas spring having a natural frequency rising with deflection, and a steel spring whose natural frequency lowers with deflection—an interesting concept. A semi-elliptic spring would locate the axle, and the hydropneumatic would carry a portion of the load.

"During the time when all GM cars used the Hotchkiss Drive, Morris was investigating the idea of a combined steel leaf spring/constant mass hydropneumatic. With this system, the front and rear suspensions would be interconnected hydraulically and a constant vibration frequency would be maintained regardless of load. It was a very good concept but never put into production."[6]

General Motors did build one of its highly-publicized "concept cars" of the late 1950s (with contemporary rocket-jet styling by Harley Earl) with hydropneumatic suspension. The GM Tech Center *Firebird* had a 1000 psi [69 bar] system for which Hirtreiter's group at Goodyear made the (fully-supported) rolling seals. These were tapered cones open at both ends, about 1.5 in. [38 mm] in diameter by two inches [51 mm] high.

Returning to Setright's account of other European developments at the time: "Finally there is perhaps the most advanced system of all—Automotive Products' Active Ride Control, which is still in the prototype stage. In effect, this is derived from Citroën's hydropneumatic design, but [has been] taken several stages further so that, in addition to self-leveling, it also virtually eliminates roll, dive, and squat, thus giving the occupants a level ride at all times. Apart from the 'amenity value,' the system also makes for improved cornering ability and tyre life by obviating the usual camber changes from roll.

"Like Citroën's, this suspension requires an engine-driven hydraulic pump and so is relatively costly to produce. It achieves its success through a highly ingenious control arrangement for the hydraulic valves. This device distinguishes between the normal road-induced wheel travel and angular movement of the body relative to the road. It opposes the latter, by varying the amount of fluid in the suspension units, without interfering with spring response over road irregularities. Consequently, the springing can be soft enough for real comfort without introducing the usual accompanying deficiencies in road behaviour. Despite the softness, roll can be eliminated (it could even be counteracted, were that desirable), and with the wheels thus immunised from camber, the road-holding and handling are vastly improved. The most recent developments of the concept have perhaps naturally, therefore, been the concern of the racing and sports-car specialist Lotus; nor is it surprising that much of this developement has involved the substitution of electronic computers for the prototype hydromechanical controls."[3]

Indeed, two decades later, it is the advent of microprocessors that seems destined to make *active suspensions* a reality in the United States. As early as 1989, General Motors' Lotus Division developed a very effective but extremely complex active suspension that used twelve sensors to measure and predict road irregularities, using hydraulics to raise and lower the corners of the vehicle to smooth out the ride. Roger Rowand reported in *Automotive News* that "Lotus' advanced active systems include computer control of a high-pressure hydraulic pump that runs high-speed hydraulic actuators that move the car's wheels up and down in response to input from sensors on the suspension and the car body. ... Lexus is using an air-suspension system with automatic electronic control of spring rate and shock absorber damping. The system senses the deflection at each wheel independently and controls pitching, bouncing, brake dive, acceleration, squat, body roll and ride height. ... Audi's newest offering, the $50,000 Audi V-8, has a three-position dial mounted on the transmission console. The driver can dial in the type of ride wanted—from 'boulevard' to 'race course.'"[7]

Despite the failure and abandonment of air suspension technology on American passenger cars two years earlier, Mercedes Benz, in 1962, released full air (all four wheels) on its 300 SE model. Commenting on the European manufacturer's move, taking the qualified position that "air suspension is a must only in Europe because of road conditions," *Automotive Industries* added cynically that "its principal benefits lie in load leveling and headlight aiming"—conceding, however, that "a properly designed air suspension system makes it possible to take advantage of softer spring rates and to provide a better ride on all manner of road conditions."[8]

The Mercedes Benz air suspension system employed three leveling valves, two in front and one at the rear. To control body roll resulting from the soft suspension, it also had *two* sway bars, one on the front and one on the rear. The system was pressurized by a single-cylinder air compressor which drew 1.3 hp from the engine. Fig. 9.3 details the front (left) and rear suspensions of the MB 300 SE, which employed 1-X (Type **B**) rolling diaphragm air springs made by the German firm, Phönix [Phoenix] Gummiwerke, a licensee of Firestone.

Two years later, in 1964, air suspension was added to the Model 600 Grand Mercedes. And in 1965, Daimler Benz introduced a hydropneumatic load-compensating spring for its cars with independent rear suspension. The unit linked the inboard ends of the swing

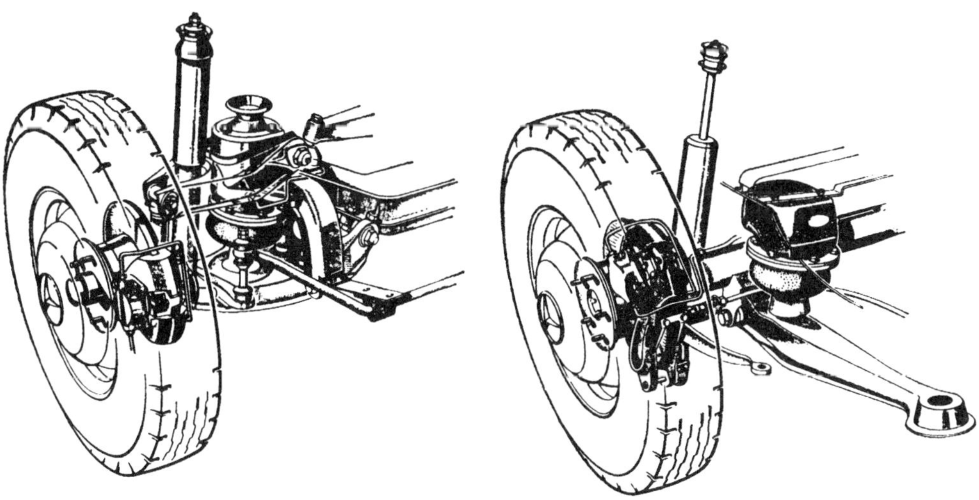

*Fig. 9.3 Front (left) and rear air suspension on the 1962 Mercedes MB 300 SE, which employed Type **B** air springs.*

axle. Although entirely automatic, its action was not unlike that of I. W. Hoagland's "bootstrap" air spring patent of 1861 (which used a miniature air spring as a pump inside the larger suspension spring). As *Automotive Industry's* European Correspondent David Scott explained, "Abnormal deflection of the axles under heavy loads effects a pumping action. This stiffens the spring, thereby restoring normal camber of the wheels and leveling the vehicle."[9]

European interest in air suspension on vehicles other than passenger cars began with rail. The first light-rail air suspension in England began with tests on London's Metropolitan Line in July, 1969. Perhaps inspired by United Aircraft's Turbotrain in the U.S., British Rail (the British Railways Board) introduced its lightweight aluminum Advanced Passenger Train in 1972, an inter-city train designed with the ambitious goal of traveling 155 mph [248 kph] on existing rails. Attaining such speeds would be due in part to the train's air suspension, which would not only provide improved isolation from rail irregularities, but also had provision for automatically tilting the coaches up to nine degrees on curves.[10]

This initial rail activity in the United Kingdom was followed in the 1970s by light-rail development in France (*Metro de Lille* and later C.I.M.T.), by the Netherlands' "Sprinter" and "Utrecht" trains, and by A.S.E.A. trains in Sweden, beginning in 1981. Early on, most of the air springs were furnished by Firestone or by its licensees, Phönix Gummiwerke and Pirelli; but European manufacturers Continental and Dunlop soon appeared to satisfy the growing market on the Continent and in the U.K.

During the 1980s, additional high-speed trains as well as light-rail London Transport commuter cars were built in England, as well as in Canada and Australia (where General Tire was also making an impact with its wide girdle band, Type **C** air spring); and the famous "Bullet Trains" appeared in Japan, their air springs made by Sumitomo—a Firestone licensee. Some of these rail developments (e.g., *Metro de Lille*) began with two-convolution bellows, but most eventually had reversible diaphragm (**B**-type) air springs in their trucks. By 1989, the *European Rubber Journal* reported that "The rolling diaphragm has taken over from the bellows in air suspensions. ... In the 1970s, the ratio of bellows to diaphragms produced was 75:25, but this is now reversed. In 1989 Dunlop made over 80,000 rolling diaphragms and nearly 27,000 bellows units."[11]

Continental AG of Hannover, Germany, is Europe's largest domestic air spring manufacturer today, making a full line of pneumatic springs, including two- and three-convolution bellows as well as 1-X (Type **B**) diaphragms and rolling lobes (Type **D**), the last of which, looking very much like their American counterpart, is shown in cross section in Fig. 9.4. Phönix remains a significant factor in Europe, especially in rail, as does Pirelli.

It was in 1977 that Scania Vabis released the first European Class 8 highway trucks on air suspension, to be followed in the '80s by Volvo, Mercedes, and M.A.N., and by Renault in 1990. Highway truck air suspension development was accelerated as a result of government studies which showed conclusively that when these heavy vehicles were equipped with air suspension they produced substantially less wear and tear on European roads. The results of the study were soon reflected in reduced license fees for trucks that rode on air.

This governmental action made a significance difference in subsequent air suspension statistics, as did the discovery that air suspension also extended *tire* life (British suspension manufacturer SAF promised increases of 70 to 150 percent in its ads). Although only 32 percent of heavy duty trucks in Europe were on air springs in 1992, this figure had risen to 65 percent by 1998. Penetration on trailers increased from 65 percent in 1992, to 82 percent by 1998. Worldwide production of truck-size air springs for 1997 was estimated at 5.5 million units. Today, all European buses and passenger rail vehicles ride on air. Rail is also 100 percent air in Japan, as are about half of Japanese buses.[12]

As this book goes to press in 1999, the new Mercedes Benz S-Class 2000 with full air suspension has just debuted. The car incorporates a number of innovations, significantly improving the ride of the car while saving 550 to 765 lb [250 to 350 kg] over recent models. Details follow in Chapter 11.

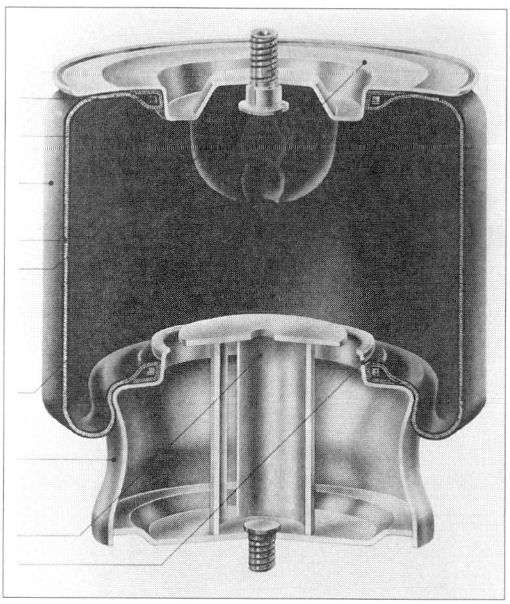

*Fig. 9.4 A Type **D** air spring made by Continental AG, Europe's largest air spring manufacturer.*

Chapter Ten

RENAISSANCE

It could be that Ford Motor Co. had heard about the passenger car air suspension activity going on at General Motors in the 1970s. Or it may simply have been, as Ford's B. K. Chance put it in the introduction to his 1984 SAE technical paper, that "The trend toward downsized vehicles and the challenge to maintain the same vehicle payload while improving overall ride quality [has] created the need for a better suspension design."[1]

General Motors activity had resurfaced briefly in the early '60s, when Goodyear's Art Hirtreiter and Tom Burkley began working with Ming Yew of the GM Tech Center to correct the infamous Corvair roll problem that had catapulted Ralph Nader to fame. Their joint effort produced an acceptable solution to the Corvair problem, but, evaluating the cost of all the related pneumatic hardware, including an engine-driven air compressor, General Motors management wisely decreed that the system was "too expensive for a cheap car."

After downsizing its "B" cars in 1977, GM chassis engineers were generally satisfied with the ride. Such was not the case when the "A" cars were downsized the following year. GM's Delco Division had recently perked up corporate interest in air suspension when, in cooperation with Goodyear, it demonstrated an air-sprung Chevette. This led to a Goodyear-Delco project that produced two full-air Pontiac Grand Prix prototypes and a Buick Regal. After the cars were extensively tested at the Milford proving ground, GM's car divisions expressed enough interest that the group looked at the Buick Riviera as well. Delco, by this time, had developed its small, 12-volt *electric* compressor, which had displaced the sometimes troublesome, vacuum-driven compressor on Cadillac's self-leveling air-shock/coil rear suspension. Operating only as needed and obviating yet another belt-driven engine appurtenance, the small, neat, Delco package did not have to be under the hood. It could bc installed in the trunk or in any other convenient location.

Prospects for air began to look favorable at GM. But the program was eventually shot down because it became obvious that one of the major defects of the earlier systems remained uncorrected. One could easily identify the air-sprung cars in the experimental garage by their various sagging attitudes if they had been parked too long. The systems leaked. Shades of the '50s. But this time there would be no coverup, and the project died.

It was Buddy K. Chance, Supervisor of Advanced Projects for Ford at the time, who was convinced that it didn't have to be this way. Electronics developments in the intervening two decades offered interesting possibilities—exemplified by Jerry W. Burns's (Delco) 1978 SAE technical paper, "A Dry Air, Electronic-Controlled Leveling System for Passenger Cars and Light Trucks,"[2] and by Toyota's concurrent development of a microprocessor-controlled suspension system that would be introduced on the 1983 Soarer.

A Goodyear team, meanwhile, had gone to Ford to demonstrate a Ford Fiesta and a mid-line Mercury that they had converted. It was fortuitous timing. Buddy Chance, who had been moved from production to his new assignment in charge of advanced engineering projects, was intrigued. His career move had been one of many initiated by Ford President Don Peterson in his effort to spark innovation in Ford Engineering at the time.

Working with Goodyear at an off-campus garage about ten miles from the Greenfield Village site of Ford Engineering, Chance's group built a full-air 1979 Ford Fairmont bristling with state-of-the-art innovation. "It rode better than the Lincoln," Chance remembers. He began campaigning for air throughout the corporation.

President Don Peterson happened to visit the experimental facility and was given a demonstration. The result would be the release of full air suspension on the 1984 Lincoln Continental and the Lincoln Mark VII.

Michael W. Soltis, of Ford's Control Systems group at the time, was brought into the program by Chance to head up electronics and to create the algorithms for the complex of automatic controls that would become the heart and brains of the system.

Compared to the way passenger car air suspension had evolved twenty years earlier, Chance's team had a refreshingly different approach. In the 1950s, the fact that air suspension had its first major success on buses was actually a disadvantage—causing it to be a poor role model for subsequent passenger car development. Buses have an inexhaustible supply of utility air, with which they operate the air brakes, open and close front and rear passenger doors, adjust air spring pressures at every stop, and sometimes perform other functions (e.g., maintaining pneumatic energy-absorbing bumpers). Their hefty compressors run all day long, continuously replenishing huge storage tanks—easily replacing any microscopic amounts of air that might leak from the air springs, from height-control valves, or from the maze of pneumatic plumbing on the 40-foot vehicle. The air cylinders that operate the doors dump much more air than such trivial amounts every time a passenger gets off the bus. Passenger car engineers tended to build a scaled-down version of this relatively crude, completely mechanical system.

To be fair, passenger car air suspension began in an age when electronics was still equated with vacuum tubes—perhaps half a dozen of which could be found in the car radio—and that was it. By contrast, the new Ford program began in an age of printed circuits, microprocessors, and personal computers having hundreds of megabytes, soon to become gigabytes, of memory.

Chance's group began by analyzing, point by point, each of the faults that had contributed to the failure of Ford's earlier release of air suspension.

"The 1958 Ford and Edsel air suspensions," Mike Soltis points out, "had 30 torque-sensitive fittings that had to be air tight; and if any one of them leaked, the whole system went down." So one obvious move would be totally eliminating threaded fittings in the air lines. But this was only the beginning. Key to success of the Ford project, would be the ability to guarantee *no* perceptible leaks from the *air springs*. A height change of less than one inch per month at the wheel was established as a criterion.

"The standard thing for a tire," Soltis explains, "is to put it under water. You look for bubbles. Using helium, and a mass spectrometer having a meter calibrated in 10^{-7} cubic centimeters per second, the threshold for a visible bubble to form is $1{,}200 \times 10^{-7}$ cc per second. Our standard was 100—an order of magnitude below that value."[3]

The air springs would be a larger-diameter version of the two-ply, beadless clamped sleeves originally developed by Goodyear for Delco, as described in Chapter 7. They would roll against hourglass pistons made of 33% glass-filled nylon.

Clamped reversible sleeve air springs, with their very thin walls, offer an advantage for passenger car applications that was not mentioned in Chapter 7. Compared with the molded-bead air springs of the 1950s, with their heavy wall gauge, they result in a reduction in short-stroke ride harshness and in noise when running over minor irregularities such as tar strips. With the very low rates characteristic of passenger car air springs (some having a rate close to zero in the nearly flat center section of their load-deflection curves), the forces introduced by such short-stroke, high-velocity incursions become noticeable as the hysteresis, or internal friction, of the rubber, which stiffens the dynamic rate at short-stroke wheel-hop frequencies. The effect is a function of the hysteresis of the elastomer selected and the thickness of the air spring wall.

Fig. 10.1 is an exploded view of the 1984 Lincoln front (and rear) air spring assemblies. The threaded piston and end cap components result in a positive compression seal around the reinforced rubber sleeve. As a part of the purchase agreement with its suppliers, Ford demands that the integrity of these seals on every air spring be verified with a "helium sniffer" leak detector before the assembly is released for shipment.

Critical to keeping the springs inflated under all conditions, regardless of what might happen in the rest of the system, is the solenoid valve at the upper right of the figure. In Fig. 10.2, the (rear and front) 1984 Lincoln air spring assemblies, this valve is the plastic projection sticking out of the end caps. Molded from 50% glass-filled nylon, the one-

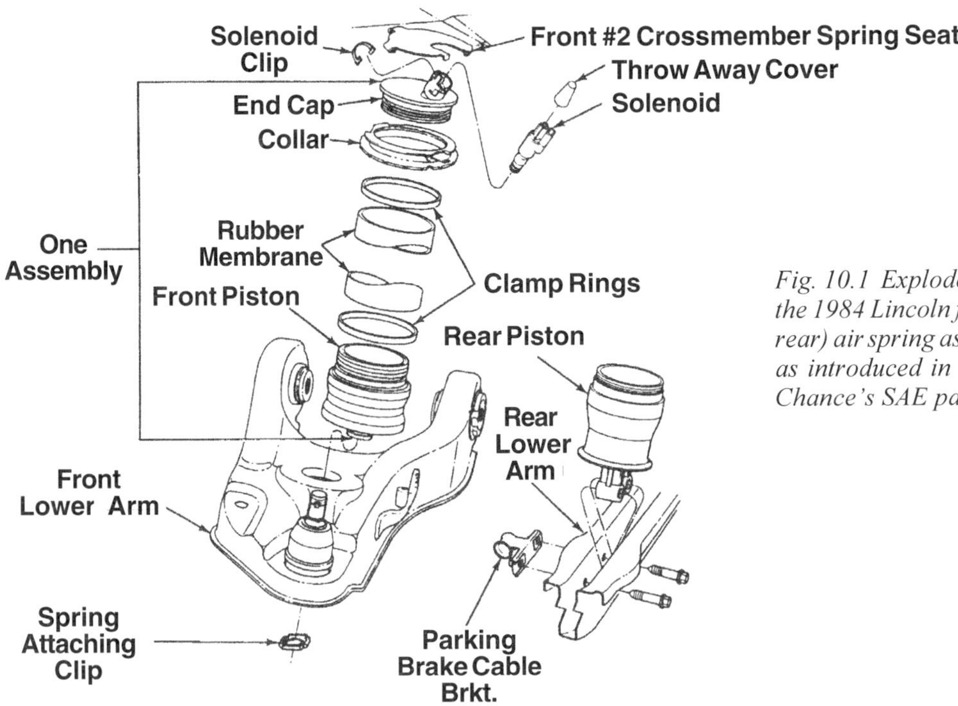

Fig. 10.1 Exploded view of the 1984 Lincoln front (and rear) air spring assemblies, as introduced in Buddy K. Chance's SAE paper.

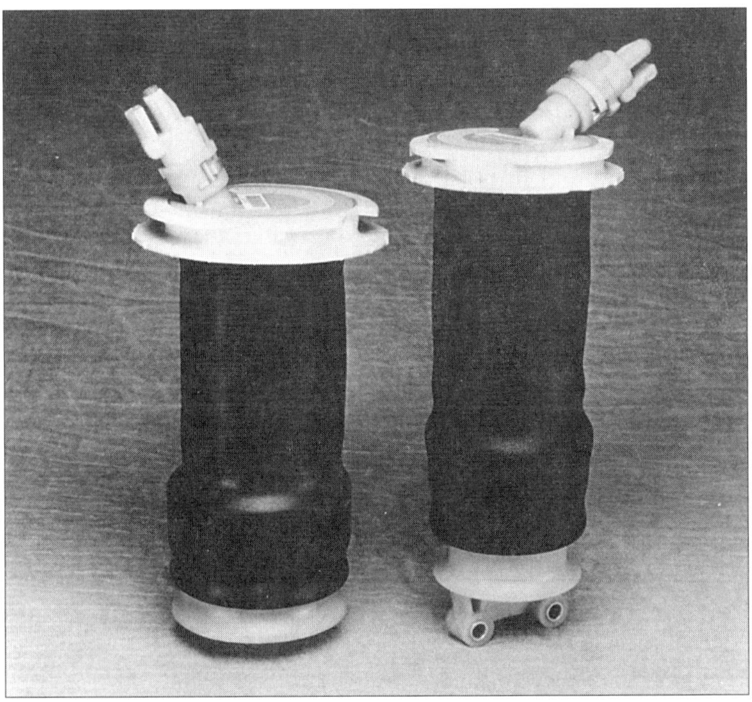

Fig. 10.2 Rear (left) and front clamped sleeve air springs developed for the 1984 Lincoln. The plastic projection is the solenoid valve, critical to maintaining reliability for keeping the springs inflated under all conditions.

piece valve terminates in separate electrical and quick-connect pneumatic fittings, as seen in the cross section in Fig. 10.3.

As Chance's 1984 SAE technical paper explains, "The valve mates directly to the air spring end cap using a double O-ring seal and a 'radiator-cap type' bayonet locking mechanism."[1] It operates like a pressure-relief valve in reverse. Pressure from *outside* the spring can blow it open. An armature spring keeps it closed, but air pressure in the spring results in additional force that drives the plunger more tightly closed. The only way air can be exhausted from the spring is for the solenoid to be energized, overcoming the combined force of the return spring and internal air pressure to lift the valve off its EPDM rubber seat. Contamination of the seal is minimized by a 40 micron filter.

Fig. 10.4 is a vintage photo of the rear air spring assembly on a 1984 Lincoln. Fig. 10.5 shows the location of the remainder of the system components. The small (25 mm bore × 10 mm stroke) single-stage air compressor is designed to require no lubrication for the life of the car. The compressor is driven by a 12-volt DC motor drawing 130 watts, similar in capacity to that of an air conditioning blower motor or a windshield wiper motor. At normal operating pressures, this air compressor delivers 14,000 cm^3 per minute [0.5 cfm].

Unlike its 1950s counterparts, the compressor has an integral *regenerative air dryer* (detailed in Fig. 10.6), which is designed to prevent the kind of freeze-ups that I frequently encountered in winter with my 1958 company Ford (and that Chevrolet attempted to prevent by introducing alcohol vapor into the system). This air dryer contains a silica gel desiccant, which absorbs moisture as high-pressure air passes from the air distribution

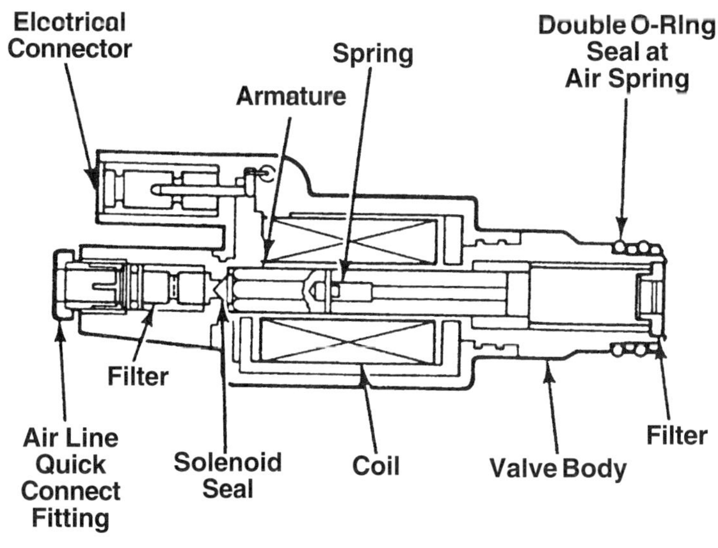

Fig. 10.3 Detail of the solenoid valve, which operates like a pressure-relief valve in reverse—requiring outside pressure from the air line to blow it ***open***.

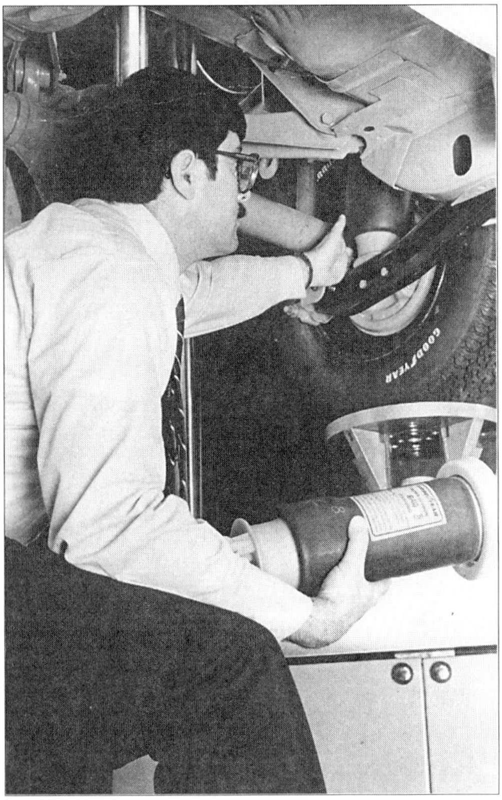

Fit. 10.4 Vintage photo of the rear air suspension on a 1984 Lincoln. (Courtesy of John Grafton's Goodyear Collection.)

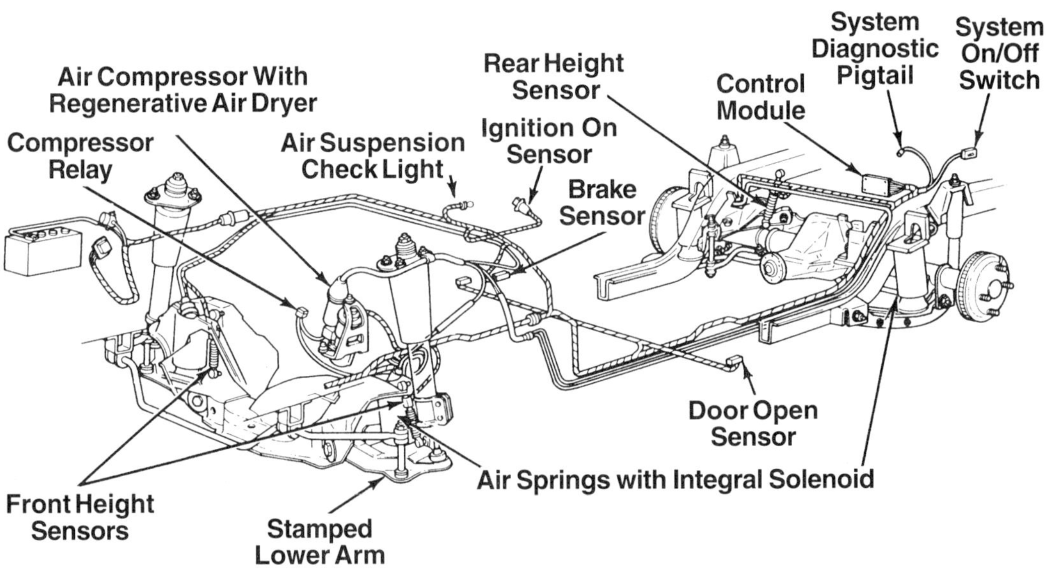

EAS System Illustration Showing the Location of Key System Components

Fig. 10.5 Components of the 1984 Lincoln air suspension system.

manifold through the dryer into the air springs (black arrows). When it is necessary to decrease spring pressure as loads change, air is vented back out through the dryer (white arrows), an action controlled by the "vent solenoid" shown. The result is *regenerative drying* of the desiccant.

As Chance's SAE paper explains, "Because most vent operations occur during engine shutdown with the air passing through the hot engine compartment, the hotter exhaust air picks up more moisture than cooler air deposits at start-up. Utilizing the pressure swing and real world temperature differences, it has been verified that the dryer provides service-free operation for the life of the vehicle."[1]

The Lincoln height-control system is a major departure from the mechanical "faucets" of the 1950s. When Mike Soltis assumed responsibility for control systems, the leveling mechanism he initially inherited—the one that had been used on the experimental Ford Fairmonts—was a rotary sensor with an optical shutter interposed between an LED and a phototransistor or photodiode. The device, originally a Delco development, sensed whether the vehicle was in trim, was too high, or too low. Although the concept had the advantage of being non-contacting (i.e., there was nothing to wear out), Soltis concluded that the geometry of translating vertical linear motion into rotary action at three different points on the car chassis compromised dead band tolerances—*and* the units were very expensive.

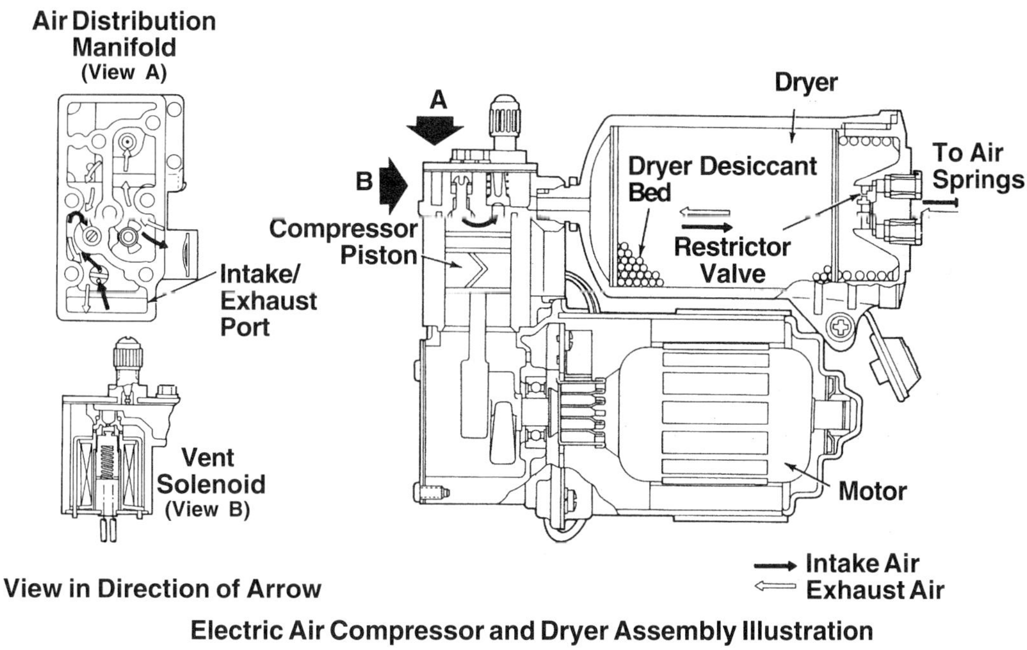

Fig. 10.6 One of many innovative features introduced on the '84 Lincoln (typical of Ford's concerted effort to avoid the pitfalls of 1950s passenger car air suspension) was a regenerative air dryer, integral with the electric compressor.

Soltis's analysis led to the development of a new *linear* height sensor costing about half as much. Shown in Fig. 10.7, the shock absorber-like configuration depends on the "Hall effect"—the induction of a potential difference in a conductor by a moving magnetic field. Chance's SAE paper describes how the sensor shaft moves up and down with vehicle suspension travel, causing two oppositely-polarized barium ferrite magnets to move past the Hall switches, inducing a pulse of current each time and identifying the direction of travel. The digital electrical signals produced are monitored by the control module of the system, which continuously counts the number of "highs" (above trim height) and subtracts the number of "lows," thus continuously monitoring vehicle ride height and making corrections as needed.

"The control module uses a 45-second averaging routine to verify that a load distribution change has actually taken place," the paper explains. "Each of the six possible leveling requests (up or down at each height sensor) are processed the same way. The sensors are read every 30 milliseconds. ... Whenever the time that a request is present exceeds the time that the request is absent by 45 seconds, the request is qualified to be serviced."[1]

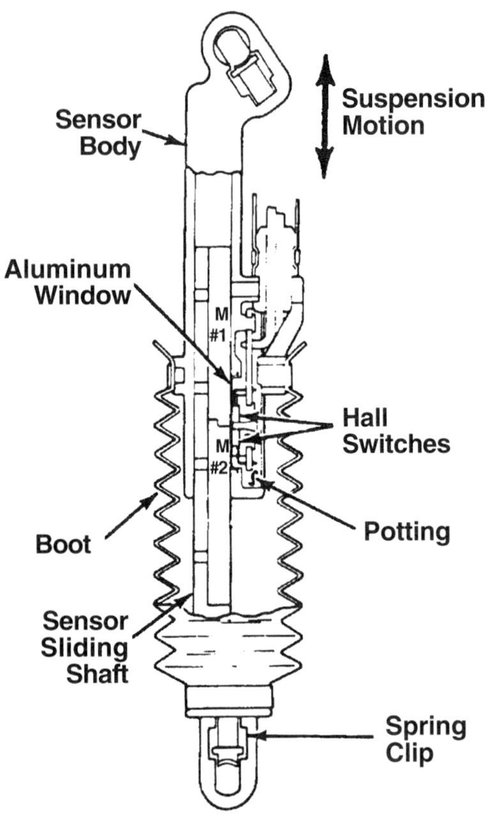

Fig. 10.7 Detail of the linear height sensor on the Lincoln, which utilizes "Hall effect" induction in combination with a microprocessor to continuously monitor and control suspension height.

The accounting of the control module is overridden by a door-open sensor, which, if there is an indication that the vehicle weight has changed, permits the system to make a correction immediately instead of waiting for the timer to run out. There is also an ignition-off mode, during which only vent requests are serviced during the first hour after the engine is switched off. At one hour, final corrections are made, but with pump corrections limited to a maximum of 30 seconds as a precaution to prevent running the battery down if the vehicle happens to be hung up on an obstacle.

This ignition-off mode was important to the new image of passenger car air suspension. In the '50s, even a system with no leaks could sag after the car was parked if the air temperature in the springs cooled down significantly. So the Lincoln system permits a single correction to be made after one hour—retrimming the vehicle to normal ride height.

Since the microcomputer could become confused by false inputs from spurious transient noise in the circuits, careful attention was given to shielding, filtering, and creating "fault-tolerant software." As an added precaution, if a repair was needed, there was a "Check Suspension" warning light in the overhead console of the car, and the control module contained a diagnostic mode that enabled a service technician to identify the problem.

I can offer a testimonial as to the effectiveness of the system. In their ten years of service during which they have accrued 80,000 miles, the air springs on my wife's 1988 Mark VII have *never* leaked any perceptible air, and she has never encountered a malfunction in the air suspension system.

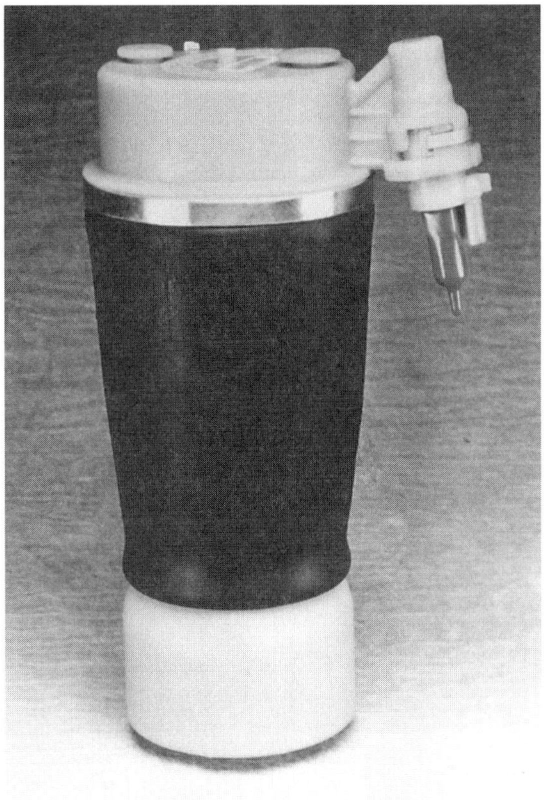

Fig. 10.8 New in 1988, a tapered air spring configuration, developed for Ford by Firestone, prevented pinching of the rubber wall when the system was deflated for any reason.

A change in the shape of the Lincoln air spring, made in 1988, is illustrated in Fig. 10.8, which depicts a rear spring assembly for the Mark VIII. The tapered sleeve prevents pinching if the springs are completely deflated.

It is perhaps ironic that, "the trend toward downsized vehicles and the challenge to maintain the same vehicle payload while improving overall ride quality" that enabled B. K. Chance to elicit corporate interest in air suspension back in 1978, ended up with the promised better suspension design on the Ford's (heaviest) flagship car. But this is consistent with the view of at least one Ford engineering executive who believes that "the future of air suspension lies in truck applications and on luxury cars."

As of this writing, air suspension is available on all Lincolns (some on the rear only), as well as on the Mercury Grand Marquis, the Ford Crown Victoria, and the Ford 4 X 4 Expedition. The variety of air spring assemblies produced for these various models appears in Fig. 10.9.

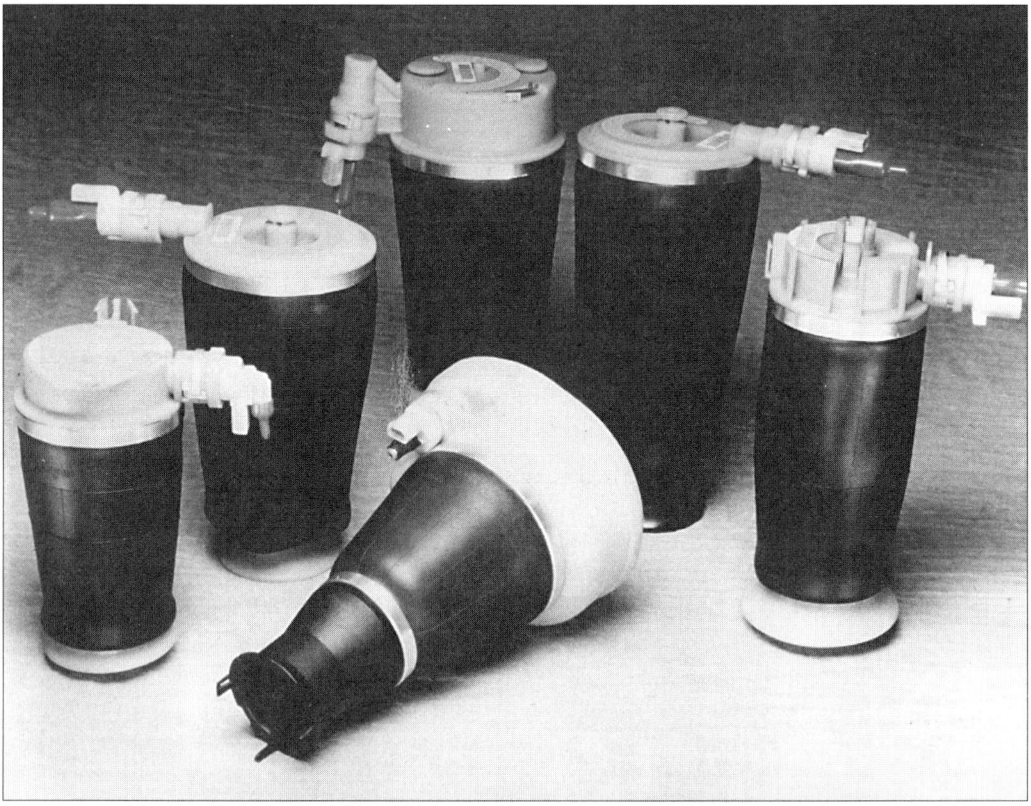

Fig. 10.9 A collection of Ford Motor Company air springs for the Lincoln, Mercury Grand Marquis, Ford Crown Victoria, and Ford 4 X 4 Expedition.

CHAPTER ELEVEN

NEXT

Enticed by the physics of the technology, engineers have occasionally pursued a number of high-tech dreams during the 50 years since pneumatic suspension became commercially viable. None has yet to become reality. One such recurring quest has been the employment of *low-gamma gases*.

The reader will remember from Chapter 4 that, in accordance with Boyle's and Charles' laws, the expression for the secondary pressure of an air spring after its volume has changed due to incremental suspension travel, under *adiabatic* (constant heat) conditions, is:

$$P_2 = P_1\left(\frac{V_1}{V_2}\right)^{1.4}$$

The exponent of 1.4 shown in this expression is the value of the *gas constant*, gamma (γ), for air. Gamma is the ratio of the specific heat of a gas at constant pressure to its specific heat at constant volume, a ratio that changes somewhat with temperature.

The value of gamma varies considerably with the size of the gas molecule—from monatomic helium (He) and argon (Ar), both of which have a gamma value of 1.66, to butane (C_4H_{10}), which has a gamma value of 1.09. Decane ($C_{10}H_{22}$) has an even lower gamma value, at 1.03, but its boiling point is higher than that of gasoline. Moreover, the properties of pneumatic springs made of rubber, or even Neoprene, would be severely degraded if they were filled with such hydrocarbons.

It is obvious from the above mathematical expression that a smaller value of this exponent will result in a smaller change in pressure. It follows that a spring filled with a low-gamma gas will have a lower dynamic rate (indeed, at the isothermal value of 1.0, some *air* springs have a *negative* rate). In 1962, Thomas Gross was issued a patent for filling "a yieldable load bearing device" with gases having a "gamma of less than 1.25: chlorofluoromethane, $CClF_3$; methyl fluoride, CHF_3; tetrafluoromethane, CF_4; perfluoroethane, CF_3–CF_3;

trifluoromethane, CHF_3; sulfur hexafluoride, SF_6; borontrifluoride, BF_3; and monobromotrifluoromethane, $CBrF_3$." Gross's patent claims name tires as an application, but they do not mention air springs.

One gas in that multisyllabic jungle, *sulfur hexafluoride* (SF_6), has promise not only because it has a gamma as low as 1.09, but because it "leaks out" only a third as fast as air. Moreover, sulfur hexafluoride is commercially available. Manufactured by Allied Signal, it is used in transformers, circuit breakers, and transmission lines because of its high dielectric strength, and also by foundries, where its chemical inertness protects the surface of magnesium castings, and where it is used to degas molten aluminum.

When Tom Bank and I discovered the possibilities offered by SF_6 we had a different objective in mind. Air springs used for vibration isolation in industrial applications either must have their air pressure checked and serviced periodically (more often than a tubeless tire because they rarely have a butyl inner liner), or, more commonly, they are connected to a compressed air line through a pressure regulator. When we were developing such applications in the 1960s, Tom and I tried filling *Airmounts* with a variety of alternative gases, equipping each of our test *Airmounts* with a visible pressure gauge. Our results seemed to confirm a hypothesis that I had entertained as a chemical engineer who had joined a tire company.

The gradual escape of air from a tire or an air spring is often cursorily attributed to the "permeability" of the elastic walls (I had heard it said that butyl, for example, has low permeability, which is why it was used for inner tubes). The word suggests, correctly, that some gas molecules somehow find their way through the organic matrix of the elastomer, driven by the higher pressure inside the part. Indeed, when Tom and I filled air springs with the small monatomic (single-atom non-molecules) of helium we found that it escaped much faster than air.

But the other mechanism causing deflation, since accepted by physicists, is that gases *dissolve* into the rubber and diffuse their way through the walls in an osmosis-like process driven by pressure differential—their solubility in the rubber depending on the chemist's truism that "like dissolve like." The four-carbon chain of butane, for example, would readily dissolve and diffuse its way through the elastomer (doing the physical properties of the air spring no favors in the process). That my hypothesis had some validity was confirmed at the time by our finding that an *Airmount* filled with carbon dioxide, CO_2. Although CO_2 is a significantly larger molecule than either the nitrogen or the oxygen of air, and although it has a higher molecular weight (38), than either N_2 (28) or O_2 (32),—the CO_2-filled spring lost its pressure much faster than one filled with air.

When I asked Dr. David Tate, then Assistant Director of Firestone Central Research, what was the largest molecule he knew of that was still a gas—and, ideally, one that was chemically dissimilar to carbon-chain elastomeric materials, he named sulfur hexafluoride.

For a gas, SF_6 has very large molecules (molecular weight 130), but, unlike large hydrocarbon molecules of similar molecular weight that are liquids at normal ambient temperatures, SF_6 has a boiling point of –63.8°C [–82.8°F]. And although we were unaware of the additional advantage at the time, it also has a very low gamma of 1.09.

This cluster of properties suggested that SF_6 could be an ideal gas for isolation applications where an air line and pressure regulator are not readily available. When Tom Bank ran leakdown tests, he confirmed that air springs filled with SF_6 retained their internal pressure several times longer than springs filled with air—*and*, as we subsequently learned, the natural frequency of the air spring is also reduced by the lower gamma. By 1988, SAE's *Manual for Incorporating Pneumatic Springs in Vehicle Suspension Designs* advised that "If sulfur hexafluoride with a γ of 1.09 is used in lieu of air with a γ of 1.38, we can expect a 20% reduction in rate."[2]

Obviously, closed-system pneumatic vehicle suspensions could be engineered with SF_6 as the spring medium. And they would require less displacement volume for a given spring rate than a system using air.

Another unorthodox method of depressing the dynamic spring rate of an air spring to soften the ride without increasing displacement volume involves literally filling the air spring with a *heat sink* that can both absorb and give off heat rapidly—effectively lowering the gamma value to something closer to isothermal unity. Ming Yew of General Motors Technical Center did considerable work on this possibility during the early days of passenger car development—which accounts for the illustration in Fig. 11.1[3] being a Type **A** reversible diaphragm air spring. The SAE HS-1576 *Pneumatic Spring Manual,* above, from which this figure is reproduced explains that, "Various materials in various forms, from aluminum balls to polyester fibers, have been used for heat sinks. ... Because metallic fibers [e.g., steel wool] are subject to corrosion and fatigue due to continuous spring deflection, their peak effectiveness is relatively short."

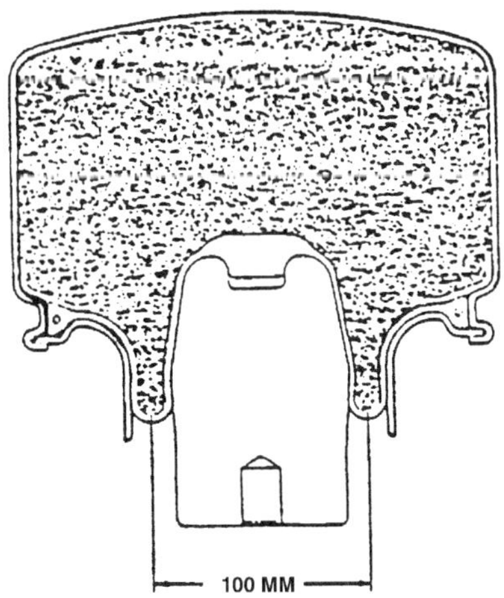

*Fig. 11.1 Type **A** air spring containing a fiber **heat sink** lowering the adiabatic rate of the spring, a research project by GM Tech Center engineer Ming Yew.*

Yew's experiments revealed that "fine polyester fibers are one of the most effective heat sinks for this type of application. Without any heat sink, the measured rate [on a full-size passenger car having a pneumatic spring suspension with a natural frequency of 1 Hz] was 14.2 N/mm. ... At the optimum point [the ideal amount of polyester fibers], the spring rate read 10.3 N/mm, a 27% reduction of pneumatic spring rate. ... It was also found that dirty or matted fibers seemed to be nearly as effective as new fibers in reducing the dynamic pneumatic spring rate." ... But, the Manual warns, "A heat sink designed to be effective for normal ride frequency will not be equally effective at high-velocity wheel impacts."[3]

One of the oldest and most elusive goals fruitlessly pursued by suspension engineers (including Roy Brown and others whose names appear earlier in this work) "since the dawn of pneumatic spring development," as the SAE *Pneumatic Spring Manual* puts it, "is *pneumatic damping*." Indeed, Firestone and Neway Equipment retraced this path as recently as ten years ago; and they probably will not be the last to discover that it is a blind alley.

"Many," the Manual editorializes, "were obsessed by the thought of partitioning the pneumatic spring chamber, and forcing the air through proper orifices to obtain the necessary damping. Unfortunately, some of their acclaimed advantages were due to increasing spring rate [as a result of having momentarily shut off displacement volume] rather than damping." The Manual presents several tables of quantified detail on the unsuccessful "Experimental Results of Air Orifice Damping."[4]

Without recycling these disappointing data, it may suffice to say that air is neither dense enough nor viscous enough, and is entirely too springy a medium to employ for hysteretic purposes. Art Hirtreiter demonstrated as much to Otto Winkelman in the 1950s (Chapter 5) when they interconnected the front and rear air suspensions of Chrysler test cars, finding that no matter how sophisticated the valving might be, they could not make air perform like hydraulic fluid.

But—having perhaps overly discounted the value of the concept—it is only fair to add that, several years ago, Mercedes debuted a van with an "air pressure proportional damping" system in which pneumatic spring pressure adjusted a piloted shock absorber valve. Not exactly air damping, but certainly related—an option that is playing a role currently in the development of active suspensions, and one that will be revisited in spades on the Mercedes S-Class 2000, as described near the end of this chapter.

In doing research for this chapter, in hopes of forecasting more likely, real-world air suspension developments yet to be realized, I spoke to a number of engineers with experience in the field, including George Villec, Senior Technical Specialist, Chassis Control Systems at Ford Motor Company's Advanced Engineering Center. Villec is intimately involved with air suspension development at Ford (and is SAE's official reviewer of this book).

Villec had an immediate response, bringing up the very practical current need for a *smaller* air spring package. Because cars with air suspension share a common "platform" (basic chassis layout, suspension hard points, and common body structure) with coil spring cars, the air spring has to fit into the same space as the coil spring for these vehicle types. Furthermore, George explains, "with the new exotic steels that are now becoming available, coil springs are getting smaller than ever!" In addition, he points out, solid axles on light trucks are being replaced by SLA (short-long arm), or wishbone suspensions providing significantly less room for the air spring, *and* the spring location is being pushed into less favorable motion ratios—being moved toward the pivot point with less vertical room for the physical stack of the air spring components.

The laws of physics being immutable, a smaller diameter spring would seem to dictate *higher pressures*. So, with the objective of developing air springs capable of operating pressures in the range of 140 psi [10 bar], rubber companies are currently experimenting with tire-cord polymers having higher modulus and higher tensile strength than nylon. Kevlar exhibits possibilities in this regard, but to date no one seems to have developed a cord dip that provides satisfactory adhesion to rubber.

Gates Rubber Company, a new entry in the air spring field, believes its "Contoured Air Spring" may provide an answer to Villec's package size problem on passenger cars. In 1990, Michael L. Crabtree of Gates was issued U.S. Patent 4,954,194 for a method of making an air spring with a *variable cord angle*. This manufacturing technique, the patent teaches, can yield an air spring having an inflated contour not unlike that of a light bulb, as shown in Fig. 11.2, one of Crabtree's patent illustrations.[5] Instead of the conventional construction of the air spring from sheets of calendered bias-cut tire cord fabric, Crabtree's technique involves building the part by winding the cords on a lathe, varying the helix angle as they wind on, thus tailoring the spring to the application. The technique may also be used, according to Gates literature, to make a tapered rolling lobe. Questions have been raised (by competitors) about the adequacy of adhesion achievable by the wind-on process compared to cord fabric which has been frictioned and skimmed on a calender.

As suggested in the figure, the design lends itself readily to installation on the straight cylinder characteristic of a McPherson strut. In operation, when the meniscus rolls into jounce, its effective diameter increases as it enters the spherical portion of the spring, increasing the effective area without requiring a flared piston to turn up the load-deflection curve near the end of the jounce stroke. Comparing this change of shape with that of a Type **A** air spring configuration (as shown in Fig. 11.1), it seems likely that their effective area vs. stroke curves would be similar on the jounce end.

Advantages claimed for the design include "reduced package size in both length and diameter, and the ability to obtain required performance with 25% to 50 % less air volume."[6] When deflected into the larger-diameter portion of its stroke, the spring is also said to present improved resistance to side loads.

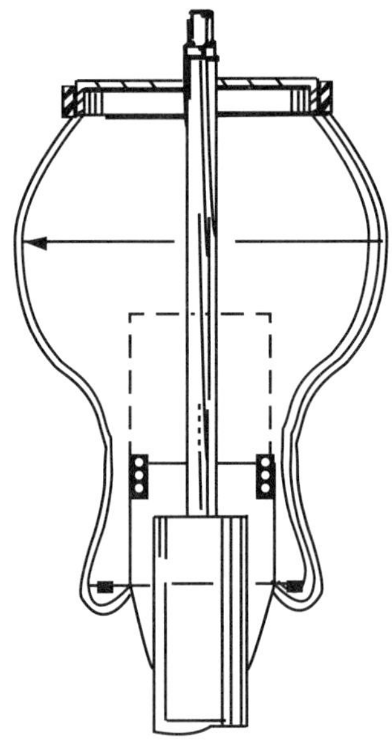

Fig. 11.2 Gates air spring, patented by Michael L. Crabtree, is built with a ***variable cord angle****. The illustration shows how the design lends itself to application on a McPherson strut.*

Michael Soltis, who was a member of Buddy Chance's team that pioneered the new air suspension for the Lincoln nearly twenty years ago, and who is now manager of his own team (AVT-Fuel Systems Technology) at Ford Motor Company's Advanced Engineering Center, believes that the future of air suspension lies primarily in truck applications, in luxury cars, and in special-purpose vehicles such as motor homes. There is work to be done on *handling*, he believes, and the focus in the development of active suspensions will be on *shock absorber technology*.

"Shock absorber control," Soltis explains, "is almost exclusively concerned with dynamics. Air suspension, for the most part, has been almost exclusively concerned with static changes. Air suspensions tend to be very closed loop at very low frequency. But active suspensions are more open loop and tend to be concerned with higher frequency stuff that is inappropriate for air." Soltis believes developers "will use the sorts of things that are associated with adjustable shock technology. Most likely, you would be taking indications from cornering, braking, and acceleration—the sort of thing that was written up in the 1984 Toyota paper on the Soarer."

The Soarer SAE technical paper, authored by Yuji Yokoya, Ken Asami, Toshimitsu Hamajima, and Noriyuki Nakashima, describes the "Toyota Electronic Modulated Suspension (TEMS) System for the 1983 Soarer." The (coil spring) passenger car was equipped with shock absorbers having adjustable orifices providing two modes, "Normal," or fairly soft, and "Sport," significantly stiffer. The system also contained sensors, switches, an electronic control unit, and actuators that turned an internal rotary valve to change the orifice size from Normal to Sport. The driver had a three-position control, marked "Normal," "Sport," and "Automatic."

When set to the Automatic mode, a microprocessor, fed by several sensors, continually adjusted the damping forces of the front and rear shock absorbers to control brake dive, squat upon acceleration, and lateral roll. The decision to switch from soft to stiff shocks was also determined by the speed of the vehicle. In controlling brake dive, for example, the shift to Sport occurred when the stop light switch was switched on, *if* the car was traveling faster than 60 km/h [37 mph] "to prevent unnecessary change of damping force when braking in city driving." The stiffer Sport setting remained in effect for two additional seconds after the stop lamp switch switched off before returning to Normal. (Early on, the developers had experimented with brake fluid pressure as a sensor, but found that the signal was "too late to restrain brake diving.")[7]

After measuring the efficacy of the system, the authors of this paper reported that their system reduced brake dive by 30 to 40 percent.

The system's anti-squat function was more complex. As the paper explains, "We found that the throttle opening can predict squat more quickly than any other displacement. In automatic transmission cars, at rapid starting, people tend to step down on the accelerator pedal more rapidly and in a greater amount. ... The damping force is changed to a greater level (Sport) on the condition that the throttle is opened from 15% to 25% or more, the throttle opening velocity is 0.2% m/sec or more, and the vehicle velocity is below 20 km/h [12 mph]. The damping force is returned to Normal on the condition that either the vehicle velocity is about 30 km/h [19 mph], or the Sport level is maintained for about 2 seconds.

"The throttle opening rate is detected by the throttle position sensor which is used in the Electronic Controlled Transmission (ECT) and other systems. The vehicle velocity is detected by the speed sensor which is used in [the] cruise control system and other electronic applications."[7]

According to the authors, the system reduced squat by 20 to 30 percent. In automatic transmission Soarers, an additional anti-squat function was provided during shifting from Neutral or Park to other positions. The function was activated when the vehicle velocity was below 10 km/h [6 mph]. The shocks returned to Normal after five seconds or when the vehicle exceeded 15 km/h [10 mph].

Anti-roll control was more sophisticated. In this regard, the authors "found that the steering wheel angle can be detected earlier than any other signals." To measure this function, they employed a slotted rotary optical shutter whose function was very similar to that of the rotary height sensor described briefly in Chapter 10. Attached to the steering column shaft, a slotted disc rotated between an LED and a phototransistor. "The number of pulses is directly proportional to the steering wheel angle," the paper explains. "A pair of photo-interrupters are located so that there is a phase difference between their output pulses to detect the steering direction." The signal is fed into "a fully integrated one-chip micro-computer with a 16 bit programmable timer and clock generator" also having both ROM and RAM memory functions. Sensing the rapidity of the steering wheel movement while taking the velocity of the vehicle into account, the computer switched the shock absorbers from Normal to Sport when cornering sharply, holding the Sport level of damping for about two seconds afterward.

This system was said to result in a 10 to 30 percent reduction in roll angle. The authors point out, however, that "the driver might sense an even greater increase in performance because the squat, nose-dive, and roll velocities have also been reduced." And they emphasize that "The TEMS system predictively controls suspension characteristics before the vehicle attitude begins to change by detecting the signals which the driver inputs."[7]

Three years after the Soarer paper, Michael Soltis delivered an SAE paper on "The 1987 Thunderbird Turbo Coupe Programmed Ride Control (PRC) Suspension," a system developed by Ford Motor Company "to satisfy the need for a suspension that can be optimized for both handling and ride, rather than compromising between the two." While performing the same basic functions as the earlier Toyota system (reducing brake dive, acceleration squat, and lateral roll without prejudicing ride), the Ford system differed significantly in its control strategy, and it contained several innovations, e.g., "The system's [patented] microprocessor control 'learns' where the straight-ahead steering position is, allowing the system to respond to absolute steering wheel angle."[8]

The 1987 Thunderbird had adjustable shock absorbers on the rear suspension and adjustable shock struts on the front. Actuated by fast-acting rotary solenoids, rotary shutter valves inside the shocks controlled the damping rate—whose resistance in their "hard" mode was three times as stiff as the force in normal "soft" damping. The PRC

(Programmed Ride Control) shock absorbers were tuned with the goal of having "the soft valving provide the best possible ride with acceptable handling," but with the firm valving providing "the best possible handling with acceptable ride."[8]

The driver had the option of manually selecting either "Automatic" or "Firm," and when the latter was selected, a "Firm Ride" light was illuminated. This same light was programmed to flash in the event of a system malfunction.

The Ford team elected to use brake hydraulic pressure to activate the stiffening of the shocks to counter brake dive instead of the stop lamp signal used by Toyota. As Soltis explains, this was because this switch "is also sensitive to light brake applications that do not require firm shock damping." The switch to Firm, in fact, was not triggered until a hydraulic brake pressure of 400 psi [28 bar] was reached.

Anti-squat stiffening of the shocks occurred when the engine entered a high-power (acceleration) condition, triggering at 90 percent throttle or 8 psi [0.3 bar] boost in manifold pressure.

Although the steering wheel shaft on the Thunderbird had a slotted disc in combination with an LED-photodiode sensor similar to the arrangement on the Soarer, the control logic for inhibiting lateral roll was quite different. The Soarer measured how fast the steering wheel was turned, a criterion rejected by Soltis's team because such a signal reacts improperly during small, rapid steering reversals that do not result in vehicle roll, and it does not respond to the large steering angles used when driving on a winding road, even if the steering angle is not changed rapidly.

Instead, Ford's microprocessor also took vehicle *speed* into account, as well as steering wheel angle, and continuously calculated lateral *accelerations*, triggering the firm mode of the shock absorbers when such acceleration exceeded 0.35 g. The signal was also activated when the vehicle speed reached 83 mph [134 kph].

These control strategy decisions were based upon considerations laid out early in the paper: "Since shock absorbers are only able to reduce transient attitude changes, it is essential that control actions be completed as early as possible in order to be effective. By basing its control decisions upon prediction of vehicle dynamic attitude changes, PRC is able to take action during the time lag from driver control input to vehicle response."[8]

Not surprisingly, in the decade since these papers were delivered, much additional work has been accomplished. Today, instead of only two damping options, "soft" and "firm," there are systems having three to five levels of damping; and it is said that some vendors are offering as many as 16 damping nuances. Also offered are shock absorbers with a continuously variable rotary shutter controlled by a rotary solenoid and others with a linear-actuated spool valve operated by a current-proportional solenoid. With controls that provide pressure proportional damping, these become a natural match with air suspension systems, as will become apparent.

But first, to keep this history in chronological order, a high-technology development that attracted the attention of *Machine Design* as early as 1988 (and the *Journal of Applied Physics* forty years earlier) is *electrorheological (ER) fluids.* These are colloids whose viscosity changes in proportion to the strength of an electric field applied to them. When first patented by W. Winslow in 1947, they consisted of finely dispersed suspensions of starch or silica gel in mineral oil. Typical of colloids, each particle has a charge (the mutual like-charge repulsion keeps them suspended). But the particles in these fluids have both positive and negative charges on their surfaces, and these are not uniformly dispersed. This imparts polarity to the particles, and when an electric field is applied, they reorient, lining up in arrays between the high voltage electrodes—which causes the shear value of the fluid to increase.[9]

Ford's Mike Soltis finds ER fluids "impressive but not imminent." Potential problems, he says, include the need for voltages in the kilovolt range, abrasiveness of the fluid, and the incompatibility of present fluids with the operating temperature range of shock absorbers. Soltis thinks that the new shock absorbers having an electromechanical, continuously variable shutter are a more practical answer for the present.[10]

The Toyota and Ford papers discussed above were focused exclusively on passenger cars. Heavy-duty vehicles equipped with air suspension—trucks, trailers, and buses whose weight changes can be more than an order of magnitude greater than those of passenger cars—have long needed adjustable shock absorbers that can be tuned to provide soft damping at light loads and stiffer damping as loads increase. At the 1997 SAE International Truck & Bus Meeting, several European engineers, together with an American university professor, presented papers on *active suspensions* for heavy-duty vehicles.[11] One of these, by Thomas Kutsche, Mannesmann Sachs, Matthias Raulf, and Hans-Otto Becher of WABCO, was reviewed in the July, 1998 issue of SAE's *Automotive Engineering*:

"Optimum riding comfort and cargo protection must be maintained with full-, partial-, and no-loads without sacrificing driving safety and good handling characteristics. With a soft suspension, ride comfort is achieved and impacts on the payload are reduced. However, optimized handling requires increased damping that can result in drastically reduced comfort when the truck is empty or partially loaded. Therefore, a damping control system that can adapt to varying driving *and* load conditions is the ideal solution."

The author points out that manually adjustable and electronically controlled shock absorbers have been available in the upper class passenger car market for some years, and that the heavy truck market in Europe is now offering such systems combined with controlled air suspensions. In Japan, he says, coaches are equipped with a manually adjustable damping system. He goes on to describe WABCO's development of a very simple, yet fully automatic system that uses pneumatically controlled shock absorbers to achieve "intelligent damping control for commercial vehicles, buses, and trailers with air suspension, working via air spring pressures. ...

"Air lines are connected to the air spring [a line from the spring to the shock absorber] as the governing element to adjust the units to firmer or softer setting as vehicle loads change. ... The system requires neither an electronic control unit nor additional sensors. Thus, it is applicable for trailers that have no separate power supply."[12]

Other systems reviewed at the 1997 Truck & Bus Meeting involved not only primary vehicle suspensions, but also the isolation of truck cabs and driver's seats. Instead of being programmed by driver input, as in the Soarer and Thunderbird passenger car systems, these systems use road input and vehicle speed to provide "intelligent vehicle control." They employ adjustable shock absorbers, the damping restrictions of which are continuously, *dynamically* controlled by velocity-sensitive or inertia-sensitive sensors. One of the systems described was a hydropneumatic suspension operating at 1500 psi [100 bar] on an off-road military vehicle in which the continuous pumping of an active fluid into and out of the hydraulic cylinder below the gas dome was also controlled by accelerometer input—the magnitude of which signal was used as a measure of the roughness of the terrain traversed by the wheels, and a criterion for establishing the degree of damping by the shock absorbers.[11]

In 1999, Daimler Benz will begin delivering the Mercedes S-Class 2000 with a host of luxury amenities, not the least of which is its four-wheel "AIRmatic suspension combined with an adaptive damping system." The air springs are clamped reversible sleeves incorporated into front and rear spring struts above the shock absorbers (Fig. 11.3).

Reduction in ride harshness, an advantage offered by reversible sleeve air spring construction (discussed in the last chapter), is emphasized in the S-Class press kit, which zeroes in on "the drawback of conventional pneumatic suspension systems whose thick-walled rubber bellows—due to high internal friction—often do not respond with sufficient sensitivity to small exciting forces, with adverse effects on ride comfort [a reference to the effects of the hysteresis in the rubber loop of the meniscus, which become noticeable in suspensions having a very low spring rate].

"This is why the new S-Class uses innovative seamless hose-type bellows for the first time comprising not the conventional two-layer diagonal reinforcement but *a single layer* of reinforcement of polyamide threads which are aligned axially. While this extremely thin-walled rubber material requires a special guide in which to move around, it does provide significantly enhanced riding comfort. When exciting forces are small, the harshness of the new-style axial bellows is more than 40 percent lower than that of a conventional two-layer spring bellows. In addition, responses are more sensitive."[13]

The springs are built with zero bias angle. The polyamide cords are parallel to the axis, as were the nylon cords on early Gabriel hydropneumatic shock absorbers of the 1960s—a construction that led to failures in the field. (When Goodyear inherited this business,

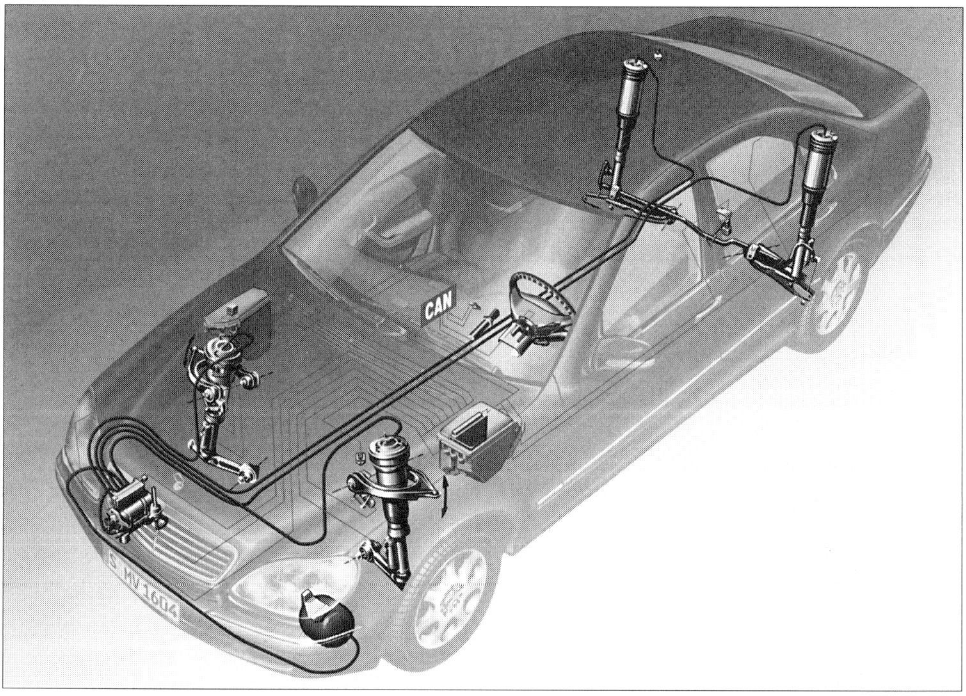

Fig. 11.3 Air springs on the 1999 Mercedes S-Class 2000 are incorporated into front and rear spring struts above the shock absorbers.

the construction became two thin plies of 10° bias angle fabric to increase cirmcumferential strength.) Mercedes precautions against the hazards Gabriel experienced include fully supporting the outside diameter of the sleeve and sealing the bottom of the metal tube with a rubber boot to exclude abrasive road dirt.

Natural frequencies of the new Mercedes air suspension are 51.6 cpm on the front, 54.6 on the rear (compared with 55.8 cpm on the front, 63.6 rear on its predecessor with coil springs).

"An electrically-driven compressor unit located under the bonnet of the new S-Class builds up the required pressure of 16 bar [230 psi] and supplies both the pneumatic suspension systems on the front and rear axles and a central accumulator positioned in the front end. A pressure sensor measures the filling pressure and ensure[s] that the values are kept constant.

"In normal driving, the spring struts are supplied directly by the compressor at a pressure of between six and ten bar [87 and 145 psi]. The compressor is switched off at speeds below ten km/h [6 mph] with pressure supplied solely from the central accumulator. The use of this dual supply system allows the saloon to be raised silently when moving off or driving at slow speed. In addition, the central accumulator is used when particularly short level control systems are required—e.g., when the car is carrying a heavy load. ...

"The pneumatic suspension system of the new S-Class not only keeps the level of the vehicle constant irrespective of the degree of loading, but also enables the driver at the press of a button on the instrument panel, to *raise the level of the vehicle.* This can be useful when driving along rough country roads or surfaces with deep potholes. In this case, the warning 'car level raised' appears in the central display on the instrument cluster."[13]

The amount of lift is much less than was offered on Buicks, Oldsmobiles, and Cadillacs in the 1950s. On Mercedes for delivery in Europe, it is 25 mm [1 in.], and on the American version it is 20 mm [0.8 in.]. Further, "If the speed of the S-Class rises to more than 80 km/h [50 mph] for a period of more than five minutes, or if a speed of 120 km/h [70 mph] is exceeded briefly, the raised level is slowly lowered back to the standard level automatically.

"On motorways, the electronically controlled AIRmatic system helps reduce wind resistance and hence fuel consumption. At speeds of 140 km/h [87 mph] or higher, the level is automatically lowered by 15 millimeters [0.6 in.]. If speed drops below 70 km/h [43 mph], the S-Class is raised back to the standard level."[13]

These speeds will not seem excessive to anyone who has driven on the German Autobahn (preferably in the hands of a professional chauffeur, who will, typically cruise at 180 km/h [112 mph] while driving a guest to his host company's headquarters). It is not surprising that specifications for Mercedes shipped to the United States have reduced limit speeds to 110 km/h [68 mph] on the high side and 60 km/h [37 mph] on the low side, respectively.

The microprocessors in the Mercedes system have the additional sophisticated assignment of "combining the pneumatic suspension with the adaptive damping system, which adapts the shock absorber rate on the front and rear axles to the vehicle load, the road surface condition, and driving style.

"When the vehicle is on the move, a steering-angle sensor, three acceleration sensors attached to the body, the speed sensor of the ABS, and the brake pedal switch constantly calculate the transverse and longitudinal accelerations of the body. The ADS control unit uses this data to generate the damper setting most appropriate for each wheel, and, within fractions of a second, issues its commands to the special switching valves located at the gas-pressure shock absorbers. They are able to set four different damper characteristics which the system selects automatically:

Stage 1: Tension soft/compression soft for comfortable ride, small body movements and low accelerations.

Stage 2: Tension soft/compression hard.

Stage 3: Tension hard/compression soft.

Stage 4: Tension hard/compression hard to minimize wheel-load fluctuations when cornering.

"Given that adjustments can be made at each wheel, it is, for example, possible selectively to raise the damping of the front wheels in order to reduce dive. A switch located in the instrument panel also allows the driver to select the standard or the firmer setting. Manual preselection of the firmer setting for a more dynamic driving style ensures that ADS will switch to Stage 4 earlier, while in standard mode switching from Stage 1 to the higher stages takes place later."[13]

It is apparent that Mercedes engineers have considered most of the issues addressed earlier in this chapter, and have probably read all of the SAE technical papers quoted in this book.

Although the American trade press has indicated that Mercedes' air suspension made it possible to reduce the weight of the S-Class 2000 by 550 lb [250 kg] on the American version and up to 765 lb [350 kg] on some European models, James Koscs of Mercedes Benz of North America says it was actually the other way around. The decision to dramatically reduce the weight of the car while still "providing a level of comfort superior even to the legendary ride offered by its predecessor" was the impetus that led to air suspension, and ultimately to the *AIRmatic* system with its host of adaptive damping controls—the same impetus that brought air suspension back to Ford in the 1980s.

Having said all of the above, and having reviewed developments in the technology that have ensued since John Lewis filed U.S. Patent 4,965 in 1847, I have also offered some modest predictions with regard to the future of air suspension: additional growth as passenger car weights decrease,[14] further technological refinement in vehicle suspensions, particularly in adjustable hydraulic damping—and active suspensions. But is that all we can expect?

Probably not—for, however accurate in discerning trends, Delphi studies never anticipate breakthroughs.[15]

NOTES

Preface

1. Edward Tunis, *Wheels*, World Publishing Co., Cleveland, Ohio, 1955; Crowell, New York, 1977.

2. Thomas A. Bank, *Vehicle Suspensions: Birth and Evolution*, private research project, 1996.

3. Roy W. Brown, "Air Springs—Tomorrow's Ride," SAE Paper No. 360109, presented at the Annual Meeting of the Society of Automotive Engineers, Detroit, Mich., 14 January 1936; *SAE Journal (Transactions)*, 1936, 126.

Chapter 1: The Medium

1. Charles Goodyear "experimented endlessly, mixing a variety of chemicals into raw rubber, achieving some success in 1837 with a patented acid and metal coating, but it wasn't until 1839 that he discovered the process of vulcanization ... patenting it in 1844." *Columbia Encyclopedia*, 5th Ed., 1993, 1108, s.v. "Goodyear, Charles."

   Although Goodyear was convinced that lead was an important ingredient in his formula, it was his addition of sulfur that finally achieved the cross-linking of the natural elastomer. But he never abandoned the lead component, and, growing ever sicker, died of lead poisoning in 1860.

2. John Lewis, "Car Spring," U.S. Patent No. 4,965, 10 February 1847. This and all other patent graphics reproduced in this book are from the air spring patent archives of the Bridgestone/Firestone Law Department, Akron, Ohio, courtesy of David A. Thomas, General Counsel.

3. *Scientific American*, May 1997, Vol. 276, No. 5, 12.

4. *Columbia Encyclopedia*, 5th Ed., 1993, 2751 s.v. "Tire." See also *The World Book Encyclopedia*, 1962, 4:309 s.v. "Dunlop, John Boyd."

   The first true bicycles, devices mechanically propelled by foot pedals or by cranks, were developed by Scottish blacksmith Kirkpatrick MacMillan in 1839, and by French inventor Ernest Michaux. These devices were "improved" upon in 1865 by Pierre Lallement, whose "high-wheeler" velocipede, which ran on iron-tired wooden rims, was known as "the boneshaker." The "safety bicycle," with wheels of approximately equal size, was first manufactured in Coventry in about 1885, by English machinist James Starley. It became popular only after the invention of the pneumatic tire by Scotsman John Boyd Dunlop in 1888. An earlier patent had been issued to another Scotsman, civil engineer Robert W. Thomson, but Thomson's tire never became a commercial success. Dunlop sold his patent in 1896 to a company that took his name, the Dunlop Tire & Rubber Co.

5. *Scientific American*, 20 July 1861, 40; I. W. Hoagland, *Carriage Spring*, U.S. Patent No. 32,848, 16 July 1861.

6. William R. Fee, "Pneumatic Spring," U.S. Patent No. 19,764, 30 March 1858; George. M. Alsop, *Carriage Spring*, U.S. Patent No. 24184, 31 May 1859.

7. John Bevan and Benjamin V. Hitchcock, "Car Spring," U.S. Patent No. 111,303, 31 January 1871.

8. W. A. Dripps, "Car Spring," U.S. Patent No. 70,177, 29 October 1867.

9. Benjamin W. Davis, "Pneumatic Bicycle Saddle," U.S. Patent No. 611,377, 27 September 1898.

10. Richard Aronstein, "Air Cushion," U.S. Patent No. 638,374, 5 December 1899.

11. George J. Bancroft, "Pneumatic Cushion for Vehicles," U.S. Patent No. 980,138, 27 December 1910.

12. Alan Sullivan, "Vehicle Wheel," U.S. Patent No. 878,754, 11 February 1908.

13. Frederick Walton, U.S. Patent No. 841,290, Jan. 15, 1907; Henry K. Bryson, U.S. Patent No. 845,516, 26 February 1907; Joseph Laus, U.S. Patent No. 877,876, 29 September 1908; Henry Mayer, U.S. Patent No. 952,012, 10 March 1910.

14. F. R. Owens, "Running Gear for Vehicles," U.S. Patent No. 858,348, 25 June 1907.

15. Warren W. Annable, "Pneumatic Spring for Vehicles," U.S. Patent 673,011, 30 April 1901.

Chapter 2: Reduction to Practice

1. A. Weiland, "Shock Absorber," U.S. Patent No. 1,579,183, 1926.

2. J. A. Heany, "Air Spring," U.S. Patent 1,694,821, 11 December 1928.

3. Benjamin Bell, "Pneumatic Spring," U.S. Patent 971,583, 4 October 1910.

4. SAE Pneumatic Spring Committee, *Manual of Principles and Application of Pneumatic Springs,* Society of Automotive Engineers, Warrendale, Pa., 1.

5. Joseph Hofmann, "Pneumatic Spring," U.S. Patents 1,077,472, 4 November 1913, and 1,094,567, 28 April 1914.

6. A. E. Forsyth, "Vehicle Suspension," U.S. Patent No. 1,977,547, 1934.

7. Herbert H. Deist, "Air Springs and Their Application to Automotive, Aircraft, and Industrial Uses," presented to Rubber & Plastics Division, ASME. New York, 4 December 1957, reprinted in *Machine Design*, 6 February 1958, 141; *Mechanical Engineering,* June 1958, 61; *Rubber World.*, July 1958, 563.

8. Jack E. Gieck, "Design With Air Springs," *Product Engineering,* 26 November 1962, 4.

9. Roy W. Brown, "Pneumatic Suspension Device," U.S. Patent No. 2,133,279, 18 October 1936. The patent has 54 claims.

10. From vintage General Motors advertising brochures (previously published in Fred L. Haushalter, *Inventors I Have Known,* Exposition Press, Inc., Jericho, N.Y., 1972, 94-5).

11. Roy W. Brown, "Air Springs—Tomorrow's Ride," SAE Paper No. 360109, presented at the Annual Meeting of the Society of Automotive Engineers, Detroit, Mich., 14 January 1936; *SAE Journal (Transactions)*, 1936, 126.

12. Personal collection of Herbert H. Deist.

13. Conversation with John G. Locklin, formerly Staff Engineer, Structures and Suspensions, GM Truck & Coach Division, August 1997.

14. H. E. Fox and D. J. LaBelle, "Motor Coach Suspensions," Society of Automotive Engineers National Transportation Meeting, Chicago, Ill., 1953; see also "Life Begins at 50 In Buses," *SAE Journal,* July 1954.

15. "Airide" is a registered trademark of the Firestone Tire & Rubber Company, now Bridgestone/Firestone, Inc.

16. These archival contributions courtesy of John Grafton's Goodyear Collection.

17. Charles O. Slemmons (General Tire & Rubber Co.), "Air Spring Suspensions in Commercial Hauling Vehicles," *SAE Journal,* July, 1954 (based on a paper presented at the SAE Metropolitan Section, New York, 10 December 1953).

18. Conversations with Donald L. Manning, formerly, Executive Engineer, Mack Trucks, Inc.; later Assistant Chief Engineer, GM Truck & Coach Division; and with Harry J. Warmkessel, formerly Manager of Bus Engineering, Mack International Motor Co., August–October, 1997.

19. Conversations with John G. Locklin, formerly Staff Engineer, Structures and Suspensions, GM Truck & Coach Division, September 1997.

20. General Motors Part No. 2292246.

21. W. G. Mayes, "Air Spring Failures on Greyhound ScenicCruisers," Firestone Central Research report to Dr. J. W. Liska, 26 October 1956.

Chapter 3: Model Trains

1. Geoffrey H. Doughty, *New York Central and Trains of the Future*, TLC Publishing, Lynchburg, Va., 1997. This chapter would not have been possible without the extensive detailed information provided in this handsome book.

2. Conversation with Geoffrey H. Doughty, 7 June 1998.

3. *Railway Age*, Simmons Boardman Publications, New York, 7 May 1956, 42 ff.

4. Conversation with Herbert H. Deist, 7 June 1998.

5. Courtesy of General Motors Electro-Motive Division, McCook, Ill.

6. Courtesy of *Railway Age*, Simmons Boardman Publications, New York.

7. Geoffrey H. Doughty, *New York Central and Trains of the Future*, TLC Publishing, Lynchburg, Va., 1997, 34 ff, 67.

Chapter 4: Complying with the Gas Laws

1. A calculation will reveal that the correct figure is 101.4 N/cm^2, but, to make it come out an even 100 N/cm^2, the bar is based on 14.5 psi.

2. Gamma is defined as the ratio of the specific heat of the gas at constant pressure to the specific heat of the gas at constant volume, or C_p/C_v. The value varies with temperature.

3. Jack E. Gieck, "Design With Air Springs", *Product Engineering,* 26 November 1962, 6–7.

4. Roy W. Brown, "Air Springs—Tomorrow's Ride," SAE Paper No. 360109, presented at the Annual Meeting of the Society of Automotive Engineers, Detroit, Mich., 14 January 1936; *SAE Journal (Transactions)*, 1936, 126.

5. Thanks to the memory of veteran suspension engineer Arthur B. Hirtreiter (Truck Engineer, Mack Trucks, Inc.; later Assistant Chief Engineer, Reo Motors; still later Chief Engineer, Suspensions and Air Springs, Goodyear Tire & Rubber Co.), conversation of October, 1997

6. Conversation with John G. Locklin, formerly Staff Engineer, Structures and Suspensions, GM Truck & Coach Division, August 1997.

Chapter 5: Mutations

1. Society of Automotive Engineers, *SAE Handbook*, Part One, SAE J511a—Pneumatic Spring Terminology, 1988, 19.08.

2. V. D. Polhemus, L. J. Kehoe Jr., F. H. Cowin, and S. L. Milliken, "Cadillac's Air Suspension for the Eldorado Brougham"—SAE Paper No. 78, presented at the National Passenger Car, Body and Materials Meeting, Detroit, Mich., 5 March 1957, Fig. 3, 7.

3. Ibid., 13 ff.

4. R. W. Perkins, "Oldsmobile New-Matic Ride," SAE Paper No. 9E, January 1958.

5. Forest McFarland, Gail Peckham, and Eric Dietrich, "The Buick Air-Poise Suspension, SAE Paper No. 9A, January 1958.

6. A. B. Hirtreiter, letters to the author, 27 July 1998 and 31 July 1998.

7. Before transistors were born, decades before the dawn of the computer age, the most sophisticated tool available to engineers and others who needed greater than slide rule accuracy was the Friden calculator—an electromechanical monster that emerged in the 1930s. Weighing 40 pounds and costing $900 in the depths of the depression, the Friden had more than a hundred keys, four sliding switches, 20 tiny knobs, and 11 thumb wheels. But all it could do was add, subtract, multiply and divide—the last very slowly and noisily, clunking and shifted its carriage from one decimal place to the next. The same set of functions with the same accuracy (with the bonus of a square root function) are available today in hand calculators costing $4.95.

8. Conversations with Joseph F. Bertsch, retired Chief Engineer, Chevrolet Engine Group, April 1998.

9. Herbert H. Deist, "Air Springs and Their Application to Automotive, Aircraft, and Industrial Uses," presented to Rubber & Plastics Division, ASME, New York, 4 December 1957; reprinted in *Machine Design*, 6 February 1958, 141; *Mechanical Engineering,* June 1958, 61; *Rubber World.*, July 1958, 563. Mr. Deist retired as Director of Advanced Process and Systems Engineering, The Firestone Tire & Rubber Company.

10. Jack E. Gieck, "Design With Air Springs," *Product Engineering,* 26 November 1962.

11. Conversation with Jack Hollis, May, 1998. Ref: Roy W. Brown and Jack L. Hollis, *Automobile Suspension*, U.S. Patent No. 3,038,716, 12 June 1962, originally filed 11 July, 1955.

12. Thomas A. Bank, *Apparatus and Method of Manufacturing Rubber Article*, U.S. Patent No. 2,959,817, 15 November 1960.

13. Kai Hansen, J. F. Bertsch, and R. E. Denzer, "1958 Chevrolet Level Air Suspension," SAE Paper No. 9D, January 1958.

14. Conversation with John T. Hoban, formerly Staff Engineer, Cadillac Motor Division, April 1998.

15. Herbert F. Rudd, "Pneumatic Cushion Structure," U.S. Patent 1,177,142, 28 March 1916.

16. Courtesy of the Goodyear Audio-Visual Department

17. C. F. O'Shea, "The Ford Approach to Air Suspension," SAE Paper No. 9C, January 1958.

18. Courtesy of Tom Burkley, retired Chief Engineer, Goodyear Air Spring Division.

19. Conversation with Thomas A. Bank, formerly Manager of Firestone Airide Development, May, 1998. Bank's recollections while recounting his conversations with Chevrolet engineers in Detroit in 1957.

20. A. B. Hirtreiter, *Rolling Lobe Type Air Spring and Method of Manufacture*, U.S. Patent 3,043,582, 10 July 1962 (filed 2 October 1956).

21. Wallace S. Berry, "The Air Coil Spring: A New Factor in Rambler Suspension," SAE Paper No. 9B, January 1958.

22. G. E. Adams, *Bedtime Stories for Young Engineers*, privately published (Adams Publishing Company), Green Valley, Ariz., 1992, 78 ff.

23. Ibid., 20.

24. Arthur B. Hirtreiter, e-mail message, 8 April 1998.

25. Elmer G. Kimmich, "Air Springs," internal Goodyear report, 28 September 1956.

Chapter 6: Wild Ride

1. Hubert Luckett, "The Truth About Air Springs," *Popular Science*, January 1958, 108 ff.

2. "Engineers Dealt Blow with Sag in Air Suspension," *Automotive News,* 28 September, 1959, 28 ff.

3. Ibid.

4. Joe H. Wherry, "First Feel Behind the Wheel, The '58 Chevrolet," *Motor Trend*, December 1957, 52.

5. Ibid. "First Feel Behind the Wheel, The '58 Ford," 29.

6. Ibid. "First Feel Behind the Wheel, The '58 Buick," 26 ff.w

7. Kai Hansen, J. F. Bertsch, and R. E. Denzer, "1958 Chevrolet Level Air Suspension," SAE Paper No. 9D, January 1958, 6.

8. Walter G. Patton, "1959 Car Design Trends," *SAE Journal*, December 1958, 34.

9. John Hoban, e-mail to the author, May 2, 1998.

Chapter 7: New Era

1. Arthur B. Hirtreiter, *Rolling Lobe Type Air Spring and Method of Manufacture*, U.S. Patent No. 3,043,582, 10 July, 1962, filed 2 October 1956.

2. Interview with Arthur B. Hirtreiter, Akron, Ohio, April, 1996.

3. Courtesy of Tom Burkley, Goodyear.

4. Goodyear Audio-Visual Department.

5. Goodyear Air Spring Design Manual.

6. Firestone Industrial Products Co.

7. Telephone conversation with John Locklin, July, 1998.

8. Goodyear News Bureau.

9. Interview with Tom Burkley, June, 1998.

10. From United Delco brochure, "Superlift," J7611, April 1965.

11. Photo by the author, courtesy of Dave Towell Cadillac, Akron, Ohio.

12. Photo by the author, courtesy of John Grafton's Goodyear collection.

13. Albert W. Foster, "A Heavy Truck Cab Suspension for Improved Ride," SAE Paper No. 780408, 27 February 27–3 March 1978.

Chapter 8: Proliferation

1. Robert F. Schmidt, "GMC Astro Aire Rear Suspension," SAE Paper No. 720106, 10–14 January 1972.

2. D. J. LaBelle, "New Commercial Vehicle Concepts Possible with Air Suspension, SAE Paper No. 570082, 4–6 November, 1957.

3. C. V. Crockett and D. J. LaBelle, "A New Concept of Light Weight Highway Tractor Design," SAE Paper No. 590281, 12–16 January 1959.

4. Edward D. Hendrickson, "History of Truck Suspension—Tandem Suspensions," SAE Paper No. 690096, 13–17 January 1969.

5. Lloyd D. Masser, "Add an Axle by the Use of an Equalizing Air Suspension," SAE Paper No. 650726, 18–21 October 1965.

6. *Commercial Car Journal*, September, 1963.

7. From Hendrickson brochure, courtesy of Tom Welch, Director of Sales, Hendrickson Trailer Suspension Systems, Canton, Ohio.

8. Photo by the author, courtesy of Daniel Ralich, Vice President, R & R Truck Sales, Akron, Ohio.

9. From Watson & Chain brochure, courtesy of Gerold Lutz, Watson & Chain Manufacturing, Inc., McKinney, Tex.

10. W. S. Locke, "IH Air Suspension," SAE Paper No. 70869, 4–6 November 1970.

11. Photos by the author, courtesy of Mary Ridenour, Hissong Kenworth, Inc., Richfield, Ohio.

12. Kenneth Stadden, "Tractor Air Suspensions," *Owner Operator,* July/ August 1987, 34 ff.; Kenneth Stadden, "Smart Spec'ing—Cab Air Suspensions," *Heavy Duty Trucking*, August 1990, 90 ff.

13. Data from Firestone Industrial Products Co., courtesy of Gary Reynolds, Director of Sales and Marketing, August, 1998.

14. Courtesy of Firestone Industrial Products Co.

15. Arthur B. Hirtreiter, letter of January 15, 1998 to Max E. Rumbaugh, Executive Vice President, SAE.

Chapter 9: The French Connection

1. Jan P. Norbye, "A Short History of Air Suspension," *Special-Interest Autos*, June-July, 1973, 20 ff.

2. Andre François, *Le Problemes de la Suspension et la Solution Hyropneumatique* (Suspension Problems and the Hydropneumatic Answer), Relations Publiques Citroën, Paris, 1981.

3. Conversations with suspension engineer Jaques Bajer, formerly of Citroën (pioneer of radial tires in the U.S.), September, 1998.

4. Joseph Lowrey, "Suspension, 1956," *The Motor*, August 29, 1956.

5. L. J. K. Setright, *Anatomy of the Automobile*, 162 ff., London, 1977, Crescent Books, New York, 1985.

6. A. B. Hirtreiter, letters to the author, 27 July 1998 and 31 July 1998.

7. Roger Rowand, "Smart Suspensions," *Automotive News OEM Edition*, E34, 15 May 1989.

8. *Automotive Industries*, 60 ff. 15 February, 1962.

9. David Scott, "European Round-Up," *Automotive Industries*, 38, 1 July 1965.

10. *Business Week*, 9 September, 1972, 46.

11. Liz White, "Trucks to Use More Air Springs," *European Rubber Journal*, Rubberex, 1990.

12. Data from Firestone Industrial Products Co., courtesy of Gary Reynolds, Director of Sales and Marketing, August, 1998.

Chapter 10: Renaissance

1. B. K. Chance, "1984 Continental Mark VII/Lincoln Continental Electronically-Controlled Air Suspension (EAS) System," SAE Paper No. 840342, 27 February, 2 March 1984.

2. Jerry W. Burns, "A Dry Air, Electronic-Controlled Leveling System for Passenger Cars and Light Trucks," SAE Paper No. 780051, 27 February, 3 March 1978.

3. Interview with Michael W. Soltis, Manager, AVT-Fuel Systems Technology Team, and George N. Villec, Senior Technical Specialist, Control Systems, Advanced Vehicle Technology, Ford Motor Company, 23 August 1998.

Chapter 11: Next

1. Thomas A. O. Gross, *Pneumatic Load Bearing Devices*, U.S. Patent No. 3,047,040, 31 July 1962.

2. SAE Spring Committee, *Manual for Incorporating Pneumatic Springs in Vehicle Suspension Designs*, HS-1576, June 1988, 25.

3. Ibid, 25 ff.

4. Ibid, 27.

5. Michael L. Crabtree, *Method for Making an Airspring and Sleeve*, U.S. Patent 4,954,194, 4 September 1990.

6. Gates Rubber Company, Denver, Colo., "Contoured Air Spring Technology."

7. Yuji Yokoya, Ken Asami, Toshimitsu Hamajima, and Noriyuki Nakashima, "Toyota Electronic Modulated Suspension (TEMS) System for the 1983 Soarer," Paper No. 84341, 27 February–7 March 1984.

8. Michael W. Soltis, "1987 Thunderbird Turbo Coupe Programmed Ride Control (PRC) Suspension," SAE Paper No. 870540, 23–27 February, 1987.

9. Theodore G. Duclos, Debra N. Acker, J. David Carlson, "Fluids That Thicken Electrically," *Machine Design*, 21 January, 1988.

10. Conversation with Michael W. Soltis, October, 1998.

11. Andrew J. Deakin, "Power Consumption in Ride of a Combat Support Vehicle Slow Active Suspension," SAE Paper No. 973205; Daniel E. Williams, Gary Lee Vandermolen, and Kenneth Brandewie, "Low Bandwidth Active Cab Suspension," SAE Paper No. 973206; Thomas Kutsche, Mannesmann Sachs, Matthias Raulf, and Hans Otto Becher, "Optimized Ride Control of Heavy Vehicles with Intelligent Suspension Control," SAE Paper No. 973207; Steven K. Karamihas Christopher B. Winkler, "Ride Performance of Heavy-Vehicle Suspensions," SAE Paper No. 973208, 18 November 1997. (Note: These papers are also available in a single SAE publication, SP-1308).

12. *Automotive Engineering International*, Truck Engineering, "Heavy Vehicle Suspension Control," July 1998, 131 ff.

13. Daimler Benz Aktiengesellschaft, "Everything About the Suspension and Control Systems," *The New Mercedes-Benz S-Class*, Media Information Press Kit, Mercedes Benz of North America, Inc., Montvale, N.J., 7 September 1998.

14. Passenger car weights may decrease dramatically due to increased use of lightweight metals such as aluminum, as well as the effects of PNGV-type programs [Partners for Next Generation Vehicles] whose goal is 80 mile-per-gallon automobiles.

15. Named for the oracular shrine of Apollo at Delphi (Greece, circa 600 b.c.) "Delphi studies" have been popular in some industries for decades. They are the product of statistical polling of industry leaders, who are asked what the future holds for their technology. This author believes they successfully forecast incremental advances, but their ability to predict genuine breakthroughs is usually no better than the mystic ravings of the entranced Apollonian priestess after inhaling sulphurous fumes issuing from a crack in the floor of the ancient temple for which they are named.

INDEX

Abbreviations are used after the page number to indicate figure (*f*), end note (*n*), or tabular material (*t*).

ABOUT THE AUTHOR

Jack Gieck is a registered Professional Engineer with a degree in chemical engineering from Iowa State University. He has been a member of the Society of Automotive Engineers (SAE) for fifty years, during which time he has contributed nine SAE technical papers—most of them introducing new products. In 1989 he was honored as an SAE Ralph R. Teetor Lecturer.

After a 34-year career in product engineering and management positions, including twelve years as an air suspension pioneer, and a stint as Firestone's first Airide Product Manager, Gieck elected early retirement as Director of New Product Development and Assistant Director of Firestone's Central Research Laboratories to become a full-time writer and producer of technical and historical nontheatrical films, founding his own company, Cinemark, Inc. Shown on the History Channel, PBS, and at many festivals, his films and videos have won more than thirty awards.

Gieck is a trustee emeritus of several historical societies and is secretary of the Akron Physics Club. This is his third book. His others include *A Photo Album of Ohio's Canal Era, 1825-1913*, Kent State University Press, and *Lichfield: The U.S. Army on Trial*, University of Akron Press. Gieck has also authored magazine articles, his most recent dealing with the physics of Jules Verne in *American Heritage of Invention and Technology*.